POGUE'S
BASICS

POGUE'S BASICS

Essential Tips and Shortcuts
(That No One Bothers to Tell You)
for Simplifying the Technology in Your Life

DAVID POGUE

FLATIRON
BOOKS
NEW YORK

www.flatironbooks.com

Production Manager: *Adriana Coada*

Library of Congress Cataloging-in-Publication Data

Pogue, David, 1963-
  Pogue's basics : essential tips and shortcuts (that no one bothers to tell you) for simplifying the technology in your life / David Pogue. — First edition.
    pages cm
  ISBN 978-1-250-05348-0 (trade paperback)
  ISBN 978-1-250-05384-8 (e-book)
  1. Technology—Miscellanea. 2. Computer science—Miscellanea. 3. Internet—Miscellanea. 4. Online social networks—Miscellanea. 5. Electronic apparatus and appliances—Miscellanea. I. Title.
  T49.P645 2014
  004—dc23

                                   2014032515

Flatiron books may be purchased for educational, business, or promotional use. For information on bulk purchases, please contact the Macmillan Corporate and Premium Sales Department at 1-800-221-7945, extension 5442, or write to specialmarkets@macmillan.com.

First Edition: January 2015

10 9 8 7 6

For Kell, Tia, and Jeff,
who remind me every day what the real basics are;
and for Nicki,
who has a way of making dreams come true

Contents

Introduction: A Therapy Session About Tech   1

Part I: Your Gadgets
Chapter 1: Phones   12
Chapter 2: Tablets   43
Chapter 3: Cameras   49
Chapter 4: Everything Else with a Plug   65

Part 2: The Computer
Chapter 5: Computers   76
Chapter 6: The Mac   111
Chapter 7: Windows   146
Chapter 8: Word Processing,
Number Crunching, Slideshowing   183
Chapter 9: What Not to Do   209

Part 3: The Internet
Chapter 10: E-mail   228
Chapter 11: Web Browsers   243
Chapter 12: Google   265
Chapter 13: Videos Online   285
Chapter 14: 12 Free Services You'll Adore   297
Chapter 15: 10 Fantastic Phone Apps
to Install Right Now   311

Part 4: Social Networks
Chapter 16: Facebook   328
Chapter 17: Twitter   335

Index   344

POGUE'S BASICS

Introduction: A Therapy Session About Tech

In this country, you need a license to do anything that might get you into trouble: drive a car; own a gun; get married.

But when it comes to technology, they shove you out of the nest without so much as a pat on the back. There's no driver's ed class for tech. There's no government pamphlet that covers the essentials. Somehow, you're just supposed to *know* how to use your camera, phone, e-book reader, GPS, printer, Web browser, e-mail system, and social network.

And that's only half of the problem.

▪ Why tech moves so fast

Every industry has a business model, right? A system of making money.

Sometimes the transaction is obvious: You hand over a bill, and you take home a jar of pickles.

Sometimes it's sneakier: You pay only $50 for an inkjet printer, but you pay many times more for the ink cartridges. (If you do the math, you'll find that printer ink comes out to about $8,000 a gallon. And you thought gas companies were greedy?)

Well, the technology industry has a little business model of its own, and it's based on *insecurity*.

Yours, to be precise.

Every year, it introduces a new version of whatever it is you've bought. That software, that phone, that tablet. Of course, nobody forces you to upgrade to the new one, but you're made to feel feeble and obsolete if you don't. "What iPhone model do you have?" "You're not still running that version of Word, are you?" "What generation iPad is that?"

This annual tech upgrade cycle has become so ingrained that the cell phone carriers invoke it with a slogan: "New every two." Every two years, you're supposed to want a new phone, a new camera, a new laptop. Otherwise, you're a loser, right?

▪ How featuritis happens

This annual insecurity cycle works out beautifully for tech companies. But there are two problems with it.

First, if we believe in "new every two," we have to *throw away* the gadgets we bought last time. They have to go somewhere. The landfills fill up with the stuff, and those discarded gadgets leak some pretty toxic juice.

Second, how do you suppose the hardware and software companies entice you to buy their new versions each year? Two words: *more features*.

Each year, they pile on more and more and more features. More controls, more buttons, more bloated software.

Trouble is, if you slather on enough new features, you eventually wind up adding stuff nobody has actually asked for. Eventually, you make your product a monstrosity.

Consider Microsoft Word: At one time, it was a word processor. Today, it's also a Web-design program, a database, and a floor wax. The average person uses about 6 percent of Microsoft Word's features (and probably feels inadequate as a result).

These companies don't seem to realize that *people have lives.* Most people have work to do, places to go, people to see, families. We're not full-time technologists.

You can see where all of this is going. Eventually, you find yourself the proud owner of technology products that overwhelm you. Maybe you know enough to get by—maybe you don't even know that—but you have the nagging suspicion that you're not getting the most out of your tech.

Who's there to teach you the features that matter and give you permission to ignore the rest?

▪ How this book happened

I first noticed the growing "I don't even know the basics" crisis years ago. I was waiting in a book publisher's office, watching a receptionist become increasingly frustrated. She was editing something on the screen. She was trying to select *just one word* in a document.

And she was doing that by painstakingly dragging her cursor across the word. Each time, she'd go a little bit too high or too low, and she'd wind up selecting an entire extra line of text.

After ten minutes of this, I couldn't stand it: "Why don't you just *double-click the word?*"

She didn't have a clue you could do that! (See page 83 if you don't, either.)

In any case, I thought I should maybe write up a list of my favorite "things you thought everybody knows, but they don't" tips on my *New York Times* blog. BOOM: 1,500 reader comments in two days.

Then I put together my favorite ten of them and demonstrated them live onstage at the 2013 TED conference (a trendy annual gathering dedicated to Technology, Entertainment, and Design). The TED people posted the video of that talk on ted.com. BOOM: 4 million views.

I was convinced that I was onto something. There really *isn't* anyone putting together a master list of the knowledge that's essential for today's technology.

Or, rather, wasn't. Congratulations: You've found it.

This book's mission in life is to collect, in one place, every essential technique you'd think everybody knows about technology—(but you'd be wrong.)

You may know some of these tips already. No problem; skim them and savor the rosy glow of smug superiority.

Other tips might seem more like geeky shortcuts. That's fine, too. Adopt the ones that seem worth it for you.

What binds all of these tips together, though, is this: They make using technology *faster and easier.* Fewer steps. Less hassle. Less annoying.

The point is that if this book teaches you only *one new trick* that makes your life easier...

Well, then it wasn't a very good book.

But you'll probably pick up a lot more than that!

▪ When to buy new tech

You'll never stop the Great Treadmill of Technology Upgrades. It's just the way of the world: Whatever technology you buy today will be obsolete soon.

But at least you can arm yourself to cope with it.

You can learn, for example, *when* new gadgets will be introduced each year, so you'll never be caught by surprise. You don't want to be that sucker who buys the iPhone 6 three weeks before the iPhone 7 comes out.

In general, Apple introduces a new iPhone every September and a new iPad every October. The iPods are updated every September, too.

Samsung's Galaxy S phones appear every April.

New camera models come out in February and October.

New everything else comes out just in time for the holidays. You can play the game in one of two ways. If you wait to buy until the end of the product's life, you'll generally save money, because the price will have slunk downward since its introduction. (The exception: Apple products. Their prices don't sink much over time.)

And if you wait to buy until the new product comes out, then you'll avoid the sinking feeling that you're a sucker.

▪ Save big money on new tech

Because we live our lives on cell phones and tablets these days, we are—finally—easing off on buying new *computers*. But when the time does come, here's a tip: Buy a *refurbished* computer. They're listed in special areas of the Web sites for Apple, Dell, HP, and so on. (To find these special pages, use Google to search, for example, for "refurbished Macs" or "refurbished Dell.")

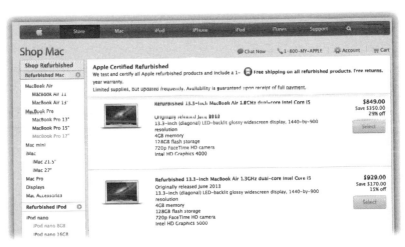

"Refurbished" isn't what it sounds like; these are brand-new computers. There's nothing wrong with them. Usually, they were bought and then returned for some reason, sometimes without even being opened. They've been inspected even more thoroughly than new machines; they have the same warranty—but they cost less.

You should also be aware that somewhere out there, at this moment, there's probably a discount coupon code for *anything* you're about to buy. Camera, blender, car wash, restaurant, flight. Before you buy anything online, search for it at RetailMeNot. com to make sure you're not throwing money away (see page 72).

▪ How to get rid of your old gadgets

Sooner or later, you'll get new electronics. Obsolescence, or your fear of being left behind, will drive you to upgrade.

Do the world a favor: Don't throw the old one in the trash.

If you think it might still have some value to someone, you can sell it to Gazelle.com, a huge online gadget recycler. Even a three-year-old used phone might get you, say, $35. And the process couldn't be easier. Gazelle sends you a box with the return postage already on it. Put in the phone, send it away, and cash the check. (It accepts even broken phones.)

If your gadget is so old or so broken that nobody would possibly want it, than drop it off at a Best Buy or Radio Shack store. Those companies offer free recycling. Just drop off your stuff and sleep well, knowing that any valuable parts of your junk will be reclaimed and reused—and that the rest will be safely disposed of.

▪ The very, very basics of technology

To use any technology, you have to know a *little* bit of terminology. This book assumes that you're familiar with a few terms and concepts.

Android: Android is software written by Google for touch-screen cell phones. It's very flexible and attractive, so it's very popular. It was designed to look and work a lot like the iPhone, but it's free to any cell phone company (Samsung, HTC, Sony, and so on) that wants to build it into its phones.

You know how the battle of computers has always been between Mac and Windows? For cell phones, it's between Android and iPhone.

You can also get tablets, computers, and even *watches* that run variations on Android.

app: It's short for *application*. When you describe your computer, you might say, "I use two programs every day: Quicken and Microsoft Word." On a phone or tablet, you'd call those *apps*.

browser: A Web browser is the program you use that enables you to read Web pages. A browser comes built into every computer, phone, and tablet—its name is probably Safari or Internet Explorer—but there are many alternative browsers, most of them free. Firefox and Chrome, for example, have amassed armies of rabid fans.

clicking: To *click* is to point the arrow cursor at something on the computer screen and then—without moving the cursor—press and release the clicker on the mouse or trackpad. To *double-click*, therefore, means to click twice in rapid succession, again without moving the cursor.

When you're told to *Shift-click* something, you click while pressing the Shift key. *Alt-clicking* and *Control-clicking* work the same way—just click while pressing the corresponding keys.

drag: To *drag* means to move the cursor while holding down the button. The purpose is usually to move something that appears on the screen, like an icon.

Fishing

My Life Story

Pizza!

DOCX

icons: These colorful inch-tall pictures represent each app, program, disk, and document on your computer, phone, or tablet. If you click an icon once, it darkens, indicating that you've just *highlighted* or *selected* it. Now you're ready to manipulate it by using, for example, a menu command.

iOS: It stands, basically, for "iPhone/iPad operating system." It's the software that runs on an iPhone, iPad, and iPod Touch.

Apple had to invent that term—iOS—because people got tired of saying, "the software that runs on an iPhone, iPad, and iPod Touch" over and over.

keyboard shortcuts: When you're at your computer, hands on the keyboard and on a roll, grabbing the mouse breaks your momentum and wastes time. That, at least, is the passionate belief of people who prefer *keyboard shortcuts:* key combinations that trigger software functions without your having to move your hands to the mouse or trackpad.

For example, in word processors, you can press Control+B to **boldface** a selected word (that's ⌘-B on the Mac).

Macs, Windows: The most popular computers in the world are Macs (made only by Apple) and Windows PCs (made by Dell, HP, Lenovo, Toshiba, and many other companies). The computer you own is probably one or the other. (Hint: If it has an Apple logo on it, it's a Mac.)

Generally, software programs—say, Photoshop or Quicken—work identically on Macs and PCs. But the keyboards on Macs and PCs are slightly different, as are the ways of doing things.

(This book has three chapters dedicated just to computers: one for Mac tips, one for Windows tips, and one for tips that work for both.)

menus: These are the words at the top of your screen: *File, Edit,* and so on. Click one to make a list of commands appear.

All right—you've now completed your orientation. Keep hands and feet inside the tram at all times; you're ready for Pogue's Tech Basics.

▪ Note About Software Versions

When you're a technology writer, you labor under a very special curse: Whatever you're writing about becomes obsolete by the time you're finished writing the book. Or even the paragraph.

Most of the tips in this book will work no matter what phone, tablet, computer, or software version you have at the moment. If you encounter some steps that don't seem to match what you're seeing, it may be that you're using an older or newer version than what's described in these pages.

Which, for the record, are these:

- **iPhone and iPad:** This book covers iOS 7 and iOS 8.

- **Android:** The version described in this book is 4.4 ("Kit-Kat"), but be warned that each phone manufacturer often makes its own changes to Android.

- **Mac:** This book describes Mac OS X 10.9 ("Mavericks") and 10.10 ("Yosemite"). Most of the illustrations show Mavericks, but despite Yosemite's cosmetic makeover, the features described on these pages work the same way.

- **Windows:** This book describes Windows 7, 8, and 8.1, with a few references to older versions tossed in for spice.

Part 1

Your Gadgets

Chapter 1: Phones

You may have heard: For the first time since the dawn of computers, sales of PCs are dropping. Fast. By like 15 percent a year.

They're being rapidly replaced by *smartphones:* beautiful, sleek, touch-screen phones that can run thousands of apps. Apps can turn a smartphone into a camera, music player, voice recorder, calendar, calculator, alarm clock, stopwatch, stock tracker, weather forecaster, flashlight, musical instrument, remote control, game machine, an e-book reader, and so on.

Some people say it can even make phone calls.

Many of the tips on the following pages direct you to adjust your phone's settings, so you need to know how to do that. On an iPhone, here's how to get there: Press the Home button (the big button below the screen). Then tap the Settings icon.

Giving instructions for Android phones is tricky, because phone companies make their own tweaks to the Android software; they put things in different places and give them different names. But in general, you make changes to Android settings like this:

Tap the Home button (the ⌂ button below the screen). Tap the Apps button (⚎). Finally, find and tap the Settings icon (*not* Google Settings, which is different).

The end-of-a-sentence automatic period trick

On a smartphone, you should complete each sentence by tapping the Space key twice.

This shortcut accomplishes three things: It creates a period, adds a space, and automatically capitalizes the next word you type. It saves you the trouble of finding the period (which, on the iPhone, is on a different keyboard layout), hitting the Space key, and then manually capitalizing the next letter.

(This technique works on every kind of smartphone: iPhone, Android, and Windows Phone. And on every BlackBerry ever made.)

Thank goodness I don't have to hunt for the period key at the end of a sentence. Or capitalize the next word.

Recharge in a hurry

You wake up. You reach for your phone. You wince: You thought that the phone had been charging all night, but it actually hadn't been plugged in right. And you have a busy day ahead of you. You have to be out the door in 30 minutes. What's the fastest possible way of charging your phone?

iPhone

First, plug it into the wall, using the little prong adapter that came with it. That'll charge it 30 minutes sooner than your computer's USB jack would.

Second, put your phone into Airplane Mode. It will charge nearly twice as quickly. (All the electricity is coming into the battery, but none is going out; the phone isn't wasting power hunting for a signal, checking e-mail, and so on.)

On the iPhone, here's how to turn on Airplane Mode: Swipe your finger up from the bottom of the screen to open the Control Center, shown at left. Tap the top-left icon.

On Android, open Settings; then, under Wireless & Networks, tap More; turn on Airplane Mode. (It might be called Flight Mode.)

Android

How to make your battery last twice as long

Having a touch-screen phone (iPhone, Android, etc.) is wonderful. You can watch movies, get driving directions, and read books.

Too bad the battery's dead by dinnertime.

But once you know which elements are using the juice, you can make each charge last far longer—a couple of days, even.

- **The screen:** It's the biggest gobbler of battery power on your phone. Turn it down to turn your battery life up.

  And how do you do that? On the iPhone, drag your finger upward from beneath the screen to make the Control Center appear. The top slider controls the screen brightness.

On an Android phone, open Settings. Tap Display, then Brightness. Turn off "Automatic brightness"; use the slider, then tap OK.

- **"Push" data:** The next biggest battery drainer is "push" e-mail, which makes new messages appear in real time. What's happening, of course, is that your phone is checking for messages *every second*, which uses power.

  On the iPhone, you can tap Settings, then Mail, Contacts, Calendars, then Fetch New Data, then turn off Push. On Android, it's Settings, E-mail Settings, Data Push.

- **Wireless features:** Your phone uses radio waves to connect to Wi-Fi hot spots and wireless Bluetooth gadgets. And a radio needs electricity.

  If you can do without Wi-Fi or Bluetooth for a while, turn those features off to save juice. On the iPhone, you'll find the on/off switches in the Control Center. On an Android phone, tap Settings; the on/off switches are right at the top.

- **Background updating:** Some apps frequently check the Internet for new information: Facebook, Twitter, stock-reporting apps, and so on, much to the dismay of your battery.

  You can turn off that feature for individual apps. On the iPhone, tap Settings, then General, then Background App Refresh. You'll see an on/off switch for each app. (There's no similar feature on Android. But some Android phones offer a feature, called Extreme Power Saver or Ultra Power Saver, that turns off background app updating—and many other features—to save energy when your charge is low.)

- **Final battery tips:** Beware of 3-D games, which can be serious power hogs. And for goodness' sake, put your phone into Airplane Mode whenever there's no cell signal—when you're in an airplane, for instance. If you forget, the phone pours even more power into its antenna, trying to find a signal—and you'll burn through it in no time.

Stop the ringing instantly

Sooner or later, it happens to everyone: Your phone starts ringing at an inopportune moment. During a movie, for example, or a wedding, or a funeral.

At that moment, you probably want to *shut the thing up, fast.* Don't be that idiot who wastes time pulling it out, waking it up, and tapping the Decline button on the screen.

Ssshh

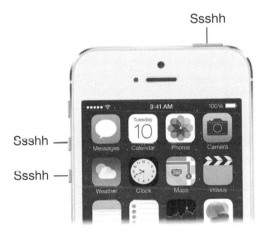

Ssshh —

Ssshh —

Instead, just *press any physical button* on the side or top. Press the power button, for example, or one of the volume keys. Often, just wrapping your fingers around the phone and squeezing hard does the trick; you'll hit one of the buttons in the process. And the phone will stop ringing.

Your caller will still hear the phone ringing sound, but the call will eventually go to voice mail.

The secret Redial button

On a cell phone, you can call back the most recent person you've called with one touch.

On a smartphone (iPhone or Android, for example), tap the Call button on the dialing pad. Doing that puts the *most recently dialed number* into the typing box, as though you'd just typed it out again. Now tap Call again to place the call.

On cell phones that have keys, the equivalent trick is pressing the Talk key when you haven't actually dialed anything yet. You get to see a list of *all* recent calls.

Bring a wet phone back from the dead

D on't beat yourself up when your phone winds up in hot water—or any kind of water. Face it: You're bringing a delicate piece of electronics into a life filled with rain, beaches, and toilets.

Phone makers are perfectly aware that more phones meet their demise from water encounters than from any other threat. That's why most cell phones contain a sticker that changes color when it gets wet; the technicians know right away how your phone really died. "Sorry, that's not covered by the warranty," they'll tell you.

But water has this delightful quality: It tends to disappear all by itself. To save a wet phone, therefore, all you have to do is make sure the evaporation happens before the damage does.

Turn off the phone. Remove and hand-dry all the pieces you can: the battery, the SIM card (the very tiny memory card that stores your account information), and memory card, for example.

Use a vacuum cleaner for 20 minutes to suck out as much water as you can. Patience, grasshopper. (Do *not* use a hair dryer, which will only blow water deeper into the phone.)

Finally, bury the phone in a container full of *uncooked rice* for 24 hours. Yes, rice. It absorbs moisture beautifully. (Change the phone's angle once an hour, if you can, to help gravity help you.) Immerse the battery, if it's removable, in a separate rice bowl.

After 24 hours, let the phone sit on a paper towel for a few hours. If there's no dampness coming from the phone, try turning it on again. You might be astonished to discover that it works just fine. (If the moisture gods are against you, take it in for repair.)

How to bypass the voice mail instructions

Your leg is on fire, or your boss is choking, or towering alien tripods are advancing upon the city. You frantically dial for help. Your call goes to voice mail. And then you have to listen to 15 seconds of *instructions*: "You may begin speaking at the tone. To page this person, press 1. When you have finished recording, you may hang up, or press 5 for more options."

Shut up. *Shut up!*

It is, in fact, possible to bypass that message with a key press, jumping directly to the "begin leaving your message" beep.

Unfortunately, to make your life as miserable as possible, each cell phone company requires a *different* keystroke to get to the beep:

- **Verizon:** Press *.

- **AT&T:** Press #.

- **Sprint:** Press 1.

- **T-Mobile:** You don't need a keystroke. Its phones don't play an instructional recording.

Of course, this means that every time you call someone, you have to know which cell phone carrier that person uses, which is a bit impractical.

If you're not sure, you can press 1, then *, then #, listening after each press. Eventually, you'll hit the right key. (The mnemonic: "One star pound.")

And if you want to do the world some good, change *your* greeting to let the world know. ("Hi, this is David. Press star to hear the beep and leave a message.")

Secrets of the three-inch keyboard

A phone with a screen that covers the entire front is great when you're watching a video, reading e-mail, or playing a game. Unfortunately, now and then, you have to *type*. That's when you long for the pleasures of a physical keyboard.

Still, that tiny on-screen keyboard isn't quite as awful as it might seem. Just be sure you realize that:

- **There doesn't seem to be a Caps Lock key**—at least not that you can see. So how are you supposed to type IN ALL CAPITALS? Simple: Tap the Shift key twice. It changes color to show that the Caps Lock key is on. (Tap once to turn it off.)

- **The keyboard gets bigger** when you turn the phone 90 degrees, becoming horizontal. Bigger keys give your fingers a bigger target.

- To type a punctuation mark, you're supposed to tap the 123 key (to display punctuation and numbers), then tap the punctuation key, then return to your typing. But you can save a step or two by *leaving your finger down* on the key and then *dragging* it onto the punctuation

Please send the lost cat to my P.O. box|
--DP

key you want. When you release your finger, your phone types the symbol *and* flips back to the alphabet keys.

The hidden pop-up punctuation keys

The letters A through Z are generally enough to get your message across. But every now and then, you might want to go to a café. And order a piña colada. And pay with a €10 bill.

Those symbols (é, ñ, €, and so on) don't appear on a smart-phone's main keyboard. You can switch to a special symbol layout—if you have all day. Fortunately, there's a great shortcut: *Hold your finger down* on a letter key to see all of its accented variations.

For example, keep your finger pressed on the A key for one second to see a pop-up menu of accented A characters (À, Á, Â, Ä, and so on). Slide onto the one you want, and marvel as your phone types it.

Not all keys sprout this pop-up palette. But the vowel keys are loaded up with diacritical marks (like ü, å, i, ő). The $ key offers a choice of other currency symbols (€, £, ¥, W). You'll find the degree symbol (°) hiding behind the letter O.

The hidden Web-address suffix keys

On the iPhone, a very similar shortcut awaits when you're trying to type a Web address or e-mail address. If you hold down your finger on the period key (.), you get a pop-up menu of common endings for Web addresses—*.com, .gov,* and *.edu.*

Use that. Learn that. It'll save you a lot of time over the years.

When autocorrect gets it autowrong

Autocorrect, of course, is that helpful phone feature that instantly fixes anything you type "incorrectly." For example, if you type *pikcle*, the phone realizes that you meant *pickle* and automatically replaces what you typed.

The problems begin when autocorrect fixes a word that was, in fact, perfectly fine. Suppose, in an effort to be cute, you type *It's very flustrating.* Right before your eyes, the phone changes

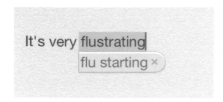

that to *It's very flu starting* (on the iPhone) or *It's very frustrating* (on Android). Which can flustrate you indeed.

Fortunately, the phone always reveals its evil plan ahead of time. On the iPhone, the replacement proposal appears in a little bubble. On Android, it's in the row of suggested words above the keyboard.

If you simply keep typing, that's the replacement you'll get. But if you can see that the suggestion is *wrong*, tap the bubble with your finger (iPhone), or tap the word you actually typed, shown first in the suggestions row (Android). That shuts the phone right up—and next time you type that word, autocorrect won't try to replace it.

(If it's too late, and you accidentally accepted the suggestion, tap the Backspace key. A word bubble appears, which you can tap to restore what you had originally typed.)

You can also turn the autocorrect feature off *entirely*. On the iPhone, tap Settings, then General, then Keyboard, then turn off Auto-Correction. On Android, it's Settings, Language & Input; tap the little settings icon next to Google Keyboard, then tap Auto-Correction and Off. (Your phone's wording may be different; Android versions vary by maker.)

What to do when "we're" auto-changes to "were"

Your smartphone is always working for you, trying to save steps and minimize annoyance. One example: It types apostrophes for you. If you tap out *dont* or *cant* or *itll*, your phone automatically types *don't* or *can't* or *it'll*.

Pretty thoughtful, eh?

But a few other words aren't so easy. If you type *were*, did you mean *we're*? If you type *ill*, did you mean *I'll*? If you type *hell*, did you mean *he'll*? About half of the time, the phone makes the wrong guess.

At least you know when it's about to guess wrong, thanks to that preview bubble. If you want a contraction and the phone doesn't realize it, *type the last letter twice.*

For instance, if you want *we're* but the phone thinks you want *were*, type an extra *e*, like this: *weree*. The phone types *we're*.

And it works with *I'll* and *she'll* and *he'll* and *we'll*, too. Type an extra *l* at the end to force the phone to create the contraction. (It doesn't work the other way. If you *don't* want the apostrophe but the phone thinks you do, you'll have to tap the suggestion bubble to reject it.)

> Well, well|
>
> Well, well|
> we'll ×
>
> Well, we'll all be home next week. |

Free directory assistance

Whatever you do, don't dial 411 on your cell phone to get directory assistance. Your cell phone carrier will slap you with a $2.50 fee for the privilege.

Instead, call 800-FREE-411 (800-373-3411). It's *free* directory assistance. The service offers both residential and business listings. You have to listen to a 10-second ad—but for most people, that's a lot more palatable than a $2.50 fee.

How to delete an app

If you can't figure out how to delete an app from your phone, it's not your fault. It's the *designer's* fault for not making it obvious.

On the iPhone, hold your finger down on any *one* app's icon until all of the icons begin to—what's the technical term?—*wiggle*.

At that point, a little X appears on each icon; tap it to delete the app.

On an Android phone, tap Settings, then Apps (or "Application manager"). In the list of apps, tap the one you want to jettison, then tap Uninstall.

In both cases, you'll discover that you're not allowed to remove certain apps. Those are the ones preinstalled by your phone's all-knowing maker (Apple or Google), which thinks it knows what's best for you.

- -

Zoom in to your iPhone

If the type on your iPhone's screen is too tiny, or you want a better view of an app, you're in luck: You can magnify the screen.

Open Settings, then General, then Accessibility. Turn on Zoom.

From now on, whenever you need a closer view, *double-tap* the screen with *three fingers at once.* You're now zoomed in. You can pan around your little Jumbotron by dragging with three fingers.

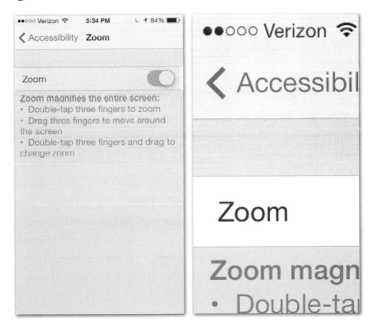

By the way, it's amazing how many people wind up triggering this feature *accidentally*. You see them lining up at Apple Stores to get their "broken" iPhones fixed! They could have saved themselves a trip—and just double-tapped with three fingers *again* to restore the regular-size screen.

- -

Take a picture of the phone screen

There are lots of reasons you might want to take a picture of your screen (called a screenshot). Maybe you're getting some infuriating error message, which you want to send to a tech-support person. Maybe you want to immortalize something you've found online (a photo, a receipt, a mention of you). Maybe you're writing a book about tech tips and you want to illustrate it.

Fortunately, it's easy.

- **On an iPhone or iPad:** Press the Home button and the Sleep (on/off) button simultaneously.

- **On an Android phone or tablet:** Press the Home button and the Volume Down button simultaneously. (Some Samsung phones require that you hold the Home and Sleep buttons instead.)

- **On a Windows 8 tablet:** While pressing the Windows logo key (⊞), press the Volume Down button.

In each case, you've created a graphics file in your Camera Roll, within your Photos app. ("Camera Roll" means "all the pictures you've taken with this gadget's own camera.") You can send it, print it, or frame it just as you would any other photo.

Overseas travel without the $6,000 phone bill

I t's no problem to take your cell phone out of the country with you. Just don't *turn it on*.

The Internet teems with stories of hapless Americans who came home from vacations to find cell phone bills for $5,000 or $6,000. In the foreign country, their phones quietly kept using the Internet in the background—checking e-mail, checking for software updates, updating Facebook posts. Unfortunately, when you're out of the country, you're subject to obscenely high international roaming rates—for Internet use, text messages, and phone calls.

You can outsmart all this, though. Here's what you need to know:

- Put your phone into Airplane Mode but turn *on* Wi-Fi. Now you can get online whenever you're in a Wi-Fi hot spot, but you'll never use the cellular network, so you'll never run up any charges.

- Call your cell phone carrier before you travel. If you sign up for its overseas-traveler plan, you can pay a one-time monthly surcharge (it's $6 for AT&T, for example) in exchange for much lower roaming rates.

- If you're on T-Mobile, don't worry. Internet use and text messages are free overseas, and phone calls to other countries are twenty cents a minute. (This free Internet service is slow Internet service—good enough for e-mail and Web sites but not for watching videos. You can pay for faster speed if you need it.)

Free maps without an Internet connection

And speaking of overseas travel: If you're a fan of the Google Maps app on your phone—and you should be—you'll like this tip. You can still use Google Maps overseas, even with your phone in Airplane Mode—*if* you've planned ahead.

While you're still home, with Internet service, call up the map area you'll want to use while you're abroad. Tap the banner at the bottom of the screen (lower left in the illustration below); on the next screen, tap "Save map to use offline" (top right).

The app gives you the opportunity to scroll the map and zoom in or out, to isolate the area you care about. (The bigger

the area, the more space the saved map will take on your phone. About five miles across is the maximum area.)

Tap Save. Type a name for your saved offline map (like "London downtown").

Once you land in London (and turned your phone to Airplane Mode to avoid going bankrupt), open Google Maps. Tap the little person icon (third from top on the facing page). Voilà: Here's the list of Offline Maps you've saved. Tap the one you want (bottom).

It opens. You can zoom in to the tiniest street details, all without having to use the Internet. The map can't give you navigation instructions, but at least you have a map to consult.

How to take better phone selfies

Every modern smartphone has a camera on both the front and the back. The one on the front (above the screen) exists so that you can join the photographic craze of the decade: selfies. (That is, self-portraits.)

Unfortunately, it's tricky to hold the camera far enough away from you to snap a decent self-portrait, especially if you're with a friend.

One solution: get a free self-timer app. There are dozens, for iPhones, Android, and so on. And in iOS 8, a timer is built into the Camera app.

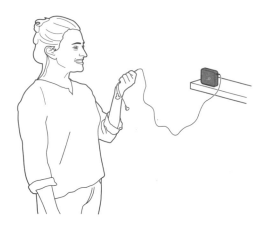

If you have an iPhone, here's another idea: plug in your white earbuds. You can use the volume-up button (+) on the cord as a remote control! Perch the iPhone on a wall, a shelf, or a barrel; then, thanks to the extra length of the earbud cord, you can back away before you snap the shot.

(You can use the middle button on the cord, the Play/Pause button, as start/stop in video mode.)

Training your gadget to stop interrupting you with hot spot names

Your phone is only trying to help.

Every time it senses that you're in a new Wi-Fi hot spot, it

interrupts whatever you're doing with an announcement like what you see at left. It's looking out for your Internet needs, and that's great. But it can get annoying if it keeps popping up while you're trying to do something on the phone—type something, dictate something, or defeat aliens.

Turns out that you can tell the phone to shut up—to *stop* offering the names of Wi-Fi hot spots it has found.

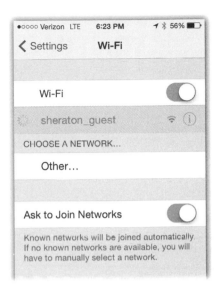

- **iPhone:** Open Settings. Tap Wi-Fi. Turn off Ask to Join Networks.

- **Android:** Open Settings. Tap "Wireless and Networks," then Wi-Fi. Turn off Network Notification.

From now on, your phone won't bug you about new Wi-Fi networks it finds. If you want to hop onto a new network, you'll have to do it manually, on the Wi-Fi page of your phone's Settings app.

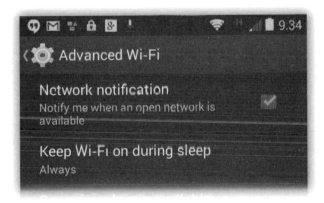

What those bizarre square bar codes are all about

They're everywhere: on cereal boxes, bus-shelter ads, magazines, brochures, airplane boarding passes. Those weird little alien-looking black-and-white squares full of square dots. Surely at some point you've wondered, "What *is* that thing?"

They're called QR codes. (It stands for Quick Response, but that won't be on the test.) Truth is, advertisers are much more excited about these codes than regular people are; most people have never scanned one in their lives.

QR codes are supposed to make it easier to get more information. When you hold up your phone to a QR code and scan it, your Web browser opens on the phone and shows you a Web page about whatever you've scanned.

That's the idea, anyway. But to scan a QR code, there's a lot of setup. First, you need a smartphone, like an iPhone or Android phone. Second, you need an *app* that reads QR codes. There are dozens of these apps, most of them free; you may as well get RedLaser (for iPhone or Android). It does a great job of reading QR codes (and traditional bar codes, too).

That's a lot of effort just to read somebody's ads. Now that you know what QR codes are, you can safely ignore them completely. That's what most people are doing already.

How to find your lost phone

Unless you're handy with duct tape, there *will* come a day when you and your cell phone are separated. Maybe you'll leave it in a hotel room; maybe you just won't remember where it is in the house. Or maybe it will slip under the seat of the car.

In all of these situations, you'll be grateful for the world's greatest free feature: Find My Phone. It shows your phone's current position on a map (if you left it somewhere in your travels)—or lets you make it start loudly chiming (so you can locate it in your house). If you're worried that some villain will riffle through your private e-mails, you can also lock the phone or even erase it, all by remote control.

Sound useful? Then set it up *now*, before you lose the phone.

- **iPhone:** Open Settings; tap iCloud. Turn on Find My iPhone. (This feature is one of the many perks of signing up for a free iCloud account, which you can do at icloud.com.)

- **Android:** Open the Google Settings app. (It's *not* the same thing as the regular Settings app.) Tap Android Device Manager. Turn on "Remotely locate this device" and "Allow remote lock and erase." Tap Activate.

Now then. When the sad day comes that you can't find your phone, here's what to do: Open up a Web browser on your computer, tablet, or another phone. Then:

- **iPhone:** Sign in to *iCloud.com*. Click Find My iPhone.

- **Android:** Sign in to *google.com/android/devicemanager*.

In either case, the Web site updates to show you, on a map, the current location of your phone. (If it's turned off or the battery is dead, you're out of luck.)

But wait, there's more. If the phone is somewhere in the house, or in some coat pocket, or in the car somewhere, you can make it start ringing loudly for a few minutes—even if it's asleep or the ringer is off.

- **iPhone:** Click the dot representing your phone, click the ⓘ next to its name, and then click Play Sound.

- **Android:** Click Ring, and then click Ring in the confirmation box.

You can also password-protect your phone by remote control.

- **iPhone:** Click Lost Mode. The Web site asks for a phone number where you can be reached; when you click Next, you can compose a message you want displayed on the iPhone's Lock screen. When you click Done, your message will appear on the phone's screen, wherever it is—and the phone will lock itself. Whoever finds it can't miss the message and can't miss the Call button that's right there on the Lock screen.

- **Android:** Click Lock. You're asked for a password that will protect your missing phone.

Finally, if there's sensitive data on the phone, you can also *erase everything on it* by remote control. (If it's ever returned, you can restore it from your backup.)

- **iPhone:** Click Lost Mode, if you haven't already. Click Erase iPhone. Confirm the dire warning box, enter your iCloud ID, and click Erase.

- **Android:** Click Erase. Click Erase in the warning box.

Note that once you've erased your phone, you can no longer use any of the Find My Phone features on it.

Dismiss the iPhone banner

Several times a day, the iPhone seems to want your attention: Every time you get an e-mail, a text message, a Facebook update, and so on. In general, such notifications appear as banners at the top of the screen.

It hangs around for a few seconds, long enough for you to read it, and then it disappears.

Flick!

You're not required to put your life on hold until it disappears, however. You can just flick it away with your finger—upward. That's handy when it's (a) blocking what you're trying to read or (b) humiliating you in front of someone you're trying to impress.

By the way: You can also shut up these notifications on an app-by-app basis. To do that, open Settings, then tap Notification Center. If you really don't need all these alerts from Facebook, for example, tap Facebook, and then tap None for the notification style.

- -

iPhone: Jump back to the top of any list

You spend much of your iPhone life in *lists*. Lists of e-mail, lists of search results, lists of photo albums. Lists of notes, of tweets, of Facebook messages.

A phone's screen is fairly small, though, at least compared with the one on your computer, TV, or local Cineplex. So you have to do a lot of scrolling.

But not when you want to jump back *up*. To return to the top of any list (in any app), *tap the top*. That is, tap the status bar above the screen. The list dutifully springs to the beginning.

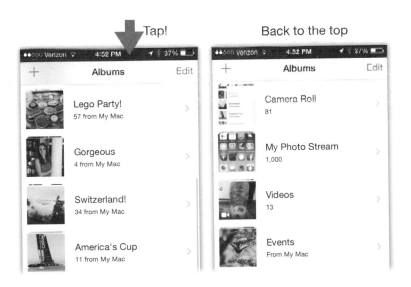

Tap!   Back to the top

Coming to an understanding with Siri

Siri is the iPhone's speech-recognition feature. Lots of people complain that her voice comprehension isn't perfect. And it's true: When you dictate a message to your iPhone (click the 🎤 button next to the Space bar on the screen), she usually makes a few transcription errors that you have to fix by hand.

That's why some people never bother "voice typing." And that's fine.

But Siri's voice *command* aspect is a different story. It virtually *always* works. Siri understands hundreds of voice commands, but here are a few especially useful ones. (Hold down the phone's Home button until you hear a double beep, then speak.)

- **"Wake me at 7:45 a.m."** You've just set your phone's alarm without having to open an app, fiddle with dials, or even touch the screen.

- **"Open Calendar."** The Open command opens any app without your having to scroll through pages of apps to find it. You can say, "Open Mail," "Open Music," "Open Safari"... anything.

- **"Call Michelle's work number."** Siri can dial any number in your phone's Contacts list. You can also say things like "Phone home," "Call Mom," or "Dial 512-444-1212." (If Siri doesn't know who your mom is, she asks you to choose your mom's card in the Contacts list. After that, Siri will always know.)

- **"What's the weather going to be like this weekend?"** The wording doesn't matter. You can also say, "Will it snow in Miami this weekend?" or "What's the high for Washington on Friday?" or "Should I wear a jacket in Fairbanks tomorrow?"

- **"What's a 15 percent tip on 37 dollars?"** Ask Siri math questions. She's amazing.

Actually, she can handle *any* factual information that would ordinarily have you skittering off to Google. "What's the capital of Germany?" "What's the world's tallest mountain?" "What flights are overhead?"

And sports trivia—oh, boy. Ask her *anything*. "What was the score of the Dolphins game last night?" "When's the next Cowboys game?" "Are the Knicks playing today?" "Who has scored the most touchdowns against the Giants?" "Who's got the highest batting average in the Major Leagues?"

Siri: She's hours of fun for the whole family.

- -

One more thing about Siri

If you utter one of the commands described in the previous tip and Siri mishears you, there's no need to start over. You can tap directly on your transcribed command and *edit* it.

You can also *scroll up* to see your earlier Siri conversations. (Most people think that once they've asked a new question, the previous questions and answers are gone forever.)

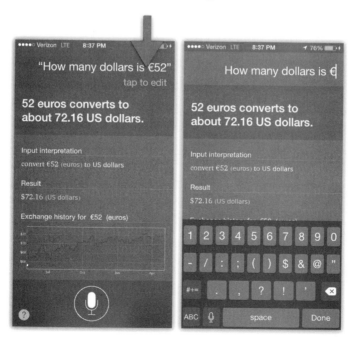

Chapter 2: Tablets

A tablet may just seem like an overgrown smartphone—and in many ways, it is. The iPad runs a variation of the same software that the iPhone runs, and an Android tablet runs a variation of the same software an Android phone runs.

In other words, almost all of the *phone* tips in the previous chapter are just as useful on your tablet. All that advice on mastering the keyboard, recharging, Find My Phone, and so on, also works on your tablet.

You're welcome.

However, a tablet's much bigger screen makes possible a realm of more intriguing software possibilities, not to mention many fewer page turns while you're reading an e-book.

Why your iPad isn't charging

Ever see a "Not charging" message on the iPad's screen even though it's plugged in? Or notice that if it's plugged in while you're using it, its battery never seems to fill up?

The explanation: An iPad's battery is like a leaky bucket. It fills up only when you pour in more juice than you're losing. There are different ways to charge an iPad—and they pour in electricity at different rates:

- **Really slow:** Charge the iPad from a low-powered USB jack, like the one on a Windows computer or a USB hub.

- **Slow:** Charge the iPad from the USB jack on a recent Mac or from an iPhone charger.

- **Fast:** Use the wall plug that came with the iPad.

And what causes those leaks in the bucket?

- **Fast leak:** Full screen brightness; playing a video game.

- **Slow leak:** Medium brightness; simple activities, like typing or answering e-mail.

- **No leak:** Turning off the iPad (putting it to sleep).

Now you understand why the iPad might not charge even when it's plugged in—if you're *using* it at the same time.

Here's another mystery solved: Sometimes when the iPad is plugged into a low-power USB jack, you might see the "Not charging" message—but it *will* charge if you put the iPad to sleep! The simple act of having the iPad *turned on* uses more electricity than it's taking in from the USB jack.

The secret iPad on-screen thumb keyboard

You know your iPad's on-screen keyboard? Unless you're an especially stretchy person, your thumbs probably aren't long enough to reach all of the keys.

But try this: Put your thumbs on the keyboard and *pull them apart.* The keyboard splits in half and shrinks. The result: a split keyboard whose keys are all thumb-reachable on both sides.

To restore the original keyboard, just push inward on both halves simultaneously.

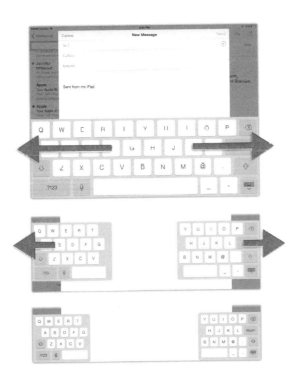

Stop your tablet screen from rotating

Your tablet's screen, as you might have noticed, always wants to be upright. If you turn the tablet in your hands, the screen image spins so that the bottom is always the bottom.

That's usually what you want. But not always.

You might, for example, sometimes like to read or watch a movie while lying down. You might want the tablet's image sideways because your *head* is sideways.

Or maybe you like to hang upside down wearing inversion boots.

In those cases, you can turn on the tablet's *rotation lock,* which stops the screen image from spinning upright all the time.

- **iPad:** Swipe upward from the bottom of the screen to reveal the Control Center. Tap the Rotation Lock button shown here.

(If you see a Mute button there instead, it's because somebody has been fiddling with the settings. Someone, possibly

you, made it so that the silencer switch—the physical switch next to the volume buttons on the side—is now the Rotation Lock switch. In this arrangement, you have to flip that switch instead.)

- **Android:** Open the Settings app; tap Display. Turn off the Auto-Rotate checkbox.

Tablet as picture frame

No matter how much you love your tablet, you probably don't use it every waking minute. (And if you do, there are some wonderful community self-help groups.)

But when you're not working, your expensive gadget doesn't have to sit there, dark and useless, as its warranty expires. You should turn it into a digital picture frame, propped up on your desk, to entertain and delight you.

- **iPad:** Open the Photos app. At the bottom of the screen, tap Albums. Tap the album that contains the pictures you want

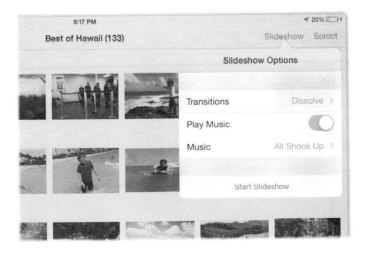

to play. Tap Slideshow at the top of the screen.

You can use the Transitions option to specify the type of crossfade you want between pictures, and the Play Music option to choose a soundtrack if desired. Then tap Start Slideshow.

(Your screen will go dark after a while to save power—unless you open Settings. Tap General; tap Auto-Lock; and choose Never.)

- **Android:** There's no built-in slideshow feature on an Android tablet. But there are dozens or hundreds of apps that are happy to perform that function for you. On your tablet, open the Google Play app—that's the app store. Search for and download a free program like Photo Slides or Slideshow Maker.

Chapter 3: Cameras

Nobody even says "digital camera" anymore. If you say "camera," everyone knows that it's a digital one. Get used to it: Film is dead. You'll have a hard time even finding film to buy these days—or places to develop it.

But that's a good thing, really. The quality of today's digital cameras is incredible. You never pay for film or processing, so you're far freer to experiment, to shoot many variations of the same shot. You become a better photographer faster.

You also get to see each photo immediately after taking it—in fact, using the screen on the back of the camera, you can see the shot *before* you take it.

And the best feature of all: Once you're digital, people can actually look at your photos. They don't end up in some box in the attic; they wind up shared by e-mail, displayed on phones, posted on Facebook, or presented on your TV.

All you have to do is learn how to *use* the damn thing.

The end of shutter lag

You know the syndrome. You're trying to photograph something that happens fast: a batter swinging, a dog jumping, your kid smiling, a diver diving.

But once you've pressed the button, there's a delay before the camera fires—and in that time, you miss the shot. You get something like the top illustration shown at left.

That waiting time—about a half a second—is called *shutter lag*. That's the interval during which the camera calculates focus and exposure. Fortunately, you can eliminate shutter lag by *prefocusing*.

To prefocus, aim the camera at the subject (if the subject hasn't arrived yet, aim at something that's the same distance from you). Half-press the shutter button. Keep your finger down. You'll hear the camera beep, meaning "I've got it!"

Now when you *fully* press, you'll get the shot *instantly*. The diver will be in the air! No shutter lag.

When not to the use the flash

Want to know when you should turn off the flash? *Whenever possible.*

At a play, a concert, or a sporting event, you should turn it off because it's useless. Your flash has a range of about eight feet; beyond that, it does nothing but make you look silly.

In any other situation, you should turn it off because flash photos look *terrible.* The light from your flash is white, harsh, and unnatural. It bleaches people's skin tone. And it turns the background into a black cave. Trouble is, if left to Automatic, most cameras tend to be flash-happy, firing in way too many situations.

So: Whenever you can turn off the flash, do. Unless it's *very* dark, you can still get a good, sharp shot if your camera has a big sensor inside (page 55) or if you've stabilized it, as described on the following page.

To force the flash off, press the lightning-bolt button. It's usually at the three o'clock position on the control dial on the back of the camera. And it displays a choice of flash settings on the screen. You want the one that says Flash Off (or the slash-circle symbol). Now the flash won't fire, no matter what.

Never take another shadowy portrait

Actually, there *is* one time when you should turn on the flash, and it might sound crazy: When you're taking pictures of people on a bright, sunny day.

Here's the problem: The camera "reads" the scene and concludes that there's tons of sunlight. But it's not smart enough to recognize that the *face you're photographing* is in shadow. You wind up with a dark, silhouetted face.

Before   After

The solution is to force the flash on—a common photographer's trick. This "fill flash" technique provides just enough light to brighten your subject's face. It eliminates the silhouette effect. Better yet, it provides flattering front light. It softens smile lines and wrinkles and puts a nice twinkle in the subject's eyes.

To force the flash, press the lightning-bolt button. This time, choose the simple lightning-bolt icon, as shown here; it may be labeled Force Flash or Flash On.

Bright sunlight: Don't fall for it!

At every graduation, wedding, and playdate on earth, you can find some amateur photographer wielding a camera and saying, "Hey, come over here in the sun so we have some light!"

If there's a professional photographer nearby, there's probably some forehead-slapping going on, too.

Direct sunlight doesn't make good portraits. It makes *terrible* portraits. It creates deep, unattractive shadows, cast by noses and eye sockets. It emphasizes wrinkles. And it makes your subject squint (below left).

Pros prefer *open shade*—go under a tree on a sunny day—or overcast skies. If you must shoot in sunlight, turn on the flash, too, as described above, and stand the subject with her back to the sun. The result is far more flattering (above right).

The only camera feature that matters

How would you feel if a salesman pressured you into buying a certain car by saying "I'm telling you, the user manual is printed on *much* sturdier paper than the other cars'!"?

Or if a real estate broker kept telling you to buy this one

house "because it's at a great elevation for ham-radio reception"?

You'd say, "Who cares!? That's irrelevant!" Right?

Well, that should be your reaction whenever you read how many *megapixels* a camera has. All that tells you is how many millions of tiny colored dots make up one of the camera's photos. It tells you nothing at all about whether that photo is *any good*. It's a cheap marketing ploy that's designed to make you think one camera's photos will look better than another's, and it's a lie.

There *is* an important statistic that really matters: the *sensor size,* as in how big the digital "film" is. The bigger the sensor, the more light the camera can absorb—the better the colors, the sharper the image, the less blur in low light.

Sensor size:

Just to make your life miserable, the camera companies don't advertise this measurement. It's not on the box. You have to go online to research it.

Even then, you have to do some math and some converting to see what you're dealing with. Small cameras' sensors are written as ratio fractions, like ½.3 of an inch. SLR camera sensors are measured in millimeters, like 16 mm x 22 mm. If you're smart, you'll use a site like *sensor-size.com* to do the conversion for you.

But you get the point: Bigger sensors are better!

The tripod in every room of your life

In low light, you run the risk of blurry photos. That's just the way it is.

It's because the shutter has to stay open long enough to soak up enough light—and while it's open, anything that moves becomes blurry. Including the camera: If the camera moves even slightly, the whole picture comes out blurry.

The trick, therefore, is to stabilize the camera. You're usually told to use a tripod. But for the average person on a trip, at a school function, or just bopping through life, carrying around a tripod is a silly suggestion.

Your first thought should be finding a big stationary object that you can use to prop the camera (or your arms): a door frame, a tree, a wall, a car, a piece of furniture.

But there's also a tripod in just

about every room in every house in the world. The threads at the top of a typical lamp—where the lampshade screws on—precisely fit the tripod mount underneath your camera. Remove the lampshade, screw the camera on, and presto: You've got a rock-steady indoor tripod. Yours free!

Make a tripod for your pocket

In their never-ending efforts to avoid having to carry a big, heavy, sharp-edged tripod around, the world's photographers have come up with another makeshift contraption that really works: the string tripod.

At the hardware store, pick up a quarter-inch steel eyehook.

Screw it into the tripod jack on the bottom of your camera. Tie a five-foot piece of string or nylon cord to it.

Then tie a weight, like a washer, to the bottom of the string.

The whole thing costs about a dollar and collapses down into something you can fit into your pocket. When you need stability, you drop the string down, stand on the far end, tug upward on the camera to keep the cord taut, and—boom. Instant steadiness!

When sharpness counts, use the self-timer

Your camera has a self-timer feature—you know, where it counts down from ten and then takes the shot automatically. You probably think of the self-timer as a feature for group photos. (You turn it on by pressing the little clock button.)

But the self-timer has another huge advantage: It lets you fire the shutter *without touching the camera.* In low light and at slow shutter speeds, even the act of pushing the shutter button is enough to jiggle the camera—and that guarantees you'll get a blurry shot.

So put the camera somewhere steady—on a table, on a car, on a tripod, on a table lamp—and let the self-timer take the picture, even if you're not in it.

You can recover photos you've already deleted

The world's camera makers know that to err is human. They've made it pretty hard to delete photos *accidentally*. You always have to confirm the deletion of a photo, sometimes more than once.

And yet every year, thousands of people still manage to delete pictures by accident that they wish they could get back.

Fortunately, your chances of retrieving deleted photos are pretty good—if you realize your mistake before you use the camera much more.

Your camera stores pictures on a memory card. Surprisingly enough, deleting photos doesn't actually delete the *data* from your memory card; it only marks the space they were occupy-

ing as now "available" for new files. (The same is true on your computer, as described on page 86.)

So if you've deleted some pictures from your camera's card, and you haven't taken many more pictures with it, get thee to the Internet. Your mission is to download a Mac or Windows program that can recover those deleted files from the card.

There are dozens of such programs. For Windows, you might try the one called Pandora Recovery (free) or Recuva (free or $25 if you want help over the phone). For the Mac, Softtote Data Recovery Free is indeed free—and worth trying first. If it doesn't seem to find your lost files, try CardRescue or Card Data Recovery. (You can run the free trial version of these apps to see how they do at finding your deleted files. For a version that can actually *recover* the deleted files, though, each app is $40.)

How to photograph streaking headlights and milky streams

You've seen this shot in a million magazines and ads: bands of colorful light streaking across a photo, formed by the headlights of passing cars. The trick is a *slow shutter speed*—keeping your camera's shutter open long enough for the cars to make some movement across the scene.

That may not be possible with a pocket cam. You really need a camera with shutter-priority mode—an SLR or an advanced pocket camera. In this mode, you can tell the camera how long to keep the shutter open—for example, a few seconds for car-taillight photos.

Stabilize your camera—tripod, wall, something. Set the shutter for four seconds. Use self-timer mode, so your finger doesn't jiggle the camera.

When you see cars coming, trip the shutter. Examine the results. If the streaks aren't long enough, add a couple of seconds to the shutter setting; if they're too long, subtract a second or two. Then publish the best ones as calendar photos.

The zoom you didn't know you had

In general, the greater the zoom power of your camera lens, the bigger the lens has to be. That's why those sports photographers sometimes use *huge* lenses that look like telescopes.

But as it turns out, you actually have more zoom than you think, thanks to a little thing called *cropping*. (This is something you do on your computer later—not on the camera.)

The modern camera takes pictures of very high resolution—that is, a lot of pixels make up the image. So many, in fact, that you can afford to shave away a lot of them and still have enough detail for a good shot.

In effect, you can zoom in by cropping into the photo on your computer (using iPhoto, Picasa, Photoshop, or whichever photo-editing program you like). Or even on your phone. It works amazingly well.

Focus and exposure lock on your phone

L ike any other camera, a smartphone makes its focusing and exposure calculations *before* it takes the picture. That's probably a good time to do it.

But what are you supposed to do when you're trying to photograph something that moves *quickly*? A track-and-field hurdle jumper? A diving-board diver? A car race? Anyone under the age of ten?

If you waited for your phone to calculate the focus and exposure at the point you tap the shutter button, you'd miss the shot.

Handily enough, your smartphone offers a feature that's usually found only on professional cameras: Auto-Exposure Lock and Autofocus Lock. This feature tells the camera to calculate its focus and exposure *now*, before the moment of truth, and lock it in. When you finally snap the shot, there won't be any delay.

To use it, point the phone at something that's the same distance from you, and has the same lighting, as the subject. For example, focus at the end of the diving board before anyone's on it.

Now:

- **iPhone:** Hold down your finger on that spot on the phone screen until you see the yellow square blink twice. When you lift your finger, the phrase "AE/AF Lock" appears.

You've now locked in exposure and autofocus. (Tap again to unlock it.) You can now snap photos, rapid-fire, without ever having to wait while your iPhone rethinks the focus and exposure.

- **Android:** Hold down your finger on the on-screen shutter button; it beeps, meaning that you've locked in focus and exposure. When you lift your finger, you'll have snapped the picture.

Chapter 4:
Everything Else
with a Plug

You're probably well aware that all kinds of things have gone digital, including books, photos, movies, TV, and phones.

But you'd be surprised at how many more of life's objects are going digital or wireless (or maybe you wouldn't be). Thermostats. Kitchen appliances. Toothbrushes. You can even buy, if you can believe it, an electronic wireless *fork*. (It notifies your smartphone and beeps if you're shoveling food into your mouth too fast.)

If you feel like it's all a little overwhelming, you're forgiven, but the pace of digital progress won't be slowing any time soon. You can either dig in and try to master it—or move to the Amish country.

Read Kindle books
without a Kindle

You know what the Kindle is, right? It's a gadget that lets you read electronic books sold by Amazon.com, which offers nearly every book imaginable as a Kindle e-book.

E-books are great. You can make the type bigger if your eyes are tired (or you're over 40). You can search for a certain word or phrase. You can get the definition of any word you don't know. And you can download any e-book in about 20 seconds instead of hauling off to a shopping mall and hoping for the best.

A Kindle, the machine that lets you read the electronic version of a book, is also great. It's portable, lightweight, and holds hundreds of books.

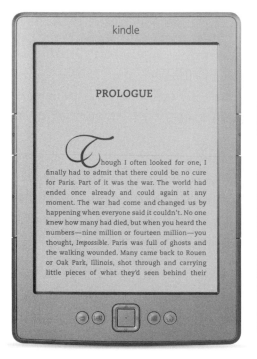

It is, however, a machine. You have to buy it, carry it around, and keep it charged.

The good news is you don't need one.

There's a free Kindle *app* for every kind of phone, tablet, and computer: iPhone, Android, Mac, Windows, and so on. In other words, you can still buy and read Kindle *books* without owning an actual Kindle *machine*—by reading them on a gadget you already own. In fact, you can read your Kindle books on the Web, wherever you happen to be; just go to *read.amazon.com*.

All of this also applies to Barnes & Noble's e-books. You don't have to buy a Nook e-book reader to read them; you can read them on the Web, on your phone, or on your tablet.

You're welcome.

- -

Why your cordless phone goes staticky

You already know that radio waves travel on different frequencies. Turn the knob, and you pull in a different station.

Those same electromagnetic waves, at different frequencies, bring you all kinds of wireless goodness: Wi-Fi Internet, cordless phones, microwave ovens.

But a problem arises when two of your wireless technologies are operating on the *same* frequency. That's why your call on a cordless phone goes staticky when someone operates the microwave, or downloads a big file, or watches an Internet video. Millions of cordless phones use the 2.4- or 5.8-gigahertz frequencies, which also happen to be the same ones that Wi-Fi hot spots use.

Your solutions are simple. Either get a cordless phone that uses a different frequency—the ones labeled DECT 6.0, for example, use the 1.9-gigahertz band—or don't use your wave-happy machines simultaneously.

The story of the third prong

America's walls are decorated with two kinds of power outlets: two-prong and three-prong. There's nothing quite as frustrating as trying to plug in a gadget that comes with a three-prong power cord—and discovering that your (old) house has only two-prong outlets! So you can't plug in!

Turns out there's some logic to this design. Of the outlet's two parallel slots, the smaller one is "hot": Electricity comes *out* of it and into your appliance. The other slot, known as "neutral," sends electricity *back into* your house. The appliance itself completes the circuit.

Any appliance with a metal case (toaster, microwave, fridge) or a metal case inside (laptop, radio) comes with a *third* prong on the cord. It goes into the hole in the wall outlet—the "ground." It's exactly the same thing as the neutral slot: an escape hatch for extra electricity.

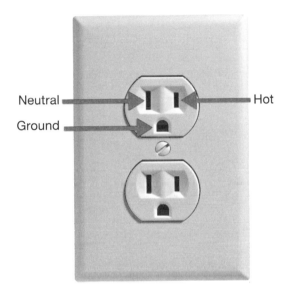

It's there for safety, for one situation only: A wire comes loose inside the appliance, sending that excess electricity into the metal case. If you touch it at that point, you could die. At the very least, you'll swear loudly.

So if you equip your gadget with one of those three-to-two adapter plugs, your gadget will work just fine. It won't know the difference. Electricity will still come out the hot slot, into the gadget, and back into the neutral slot. But you'll be defeating what engineers consider a very important safety feature.

The secrets of the Apple earbuds

Every iPad, iPod, and iPhone comes with a pair of Apple's famous white earbuds. Most people don't do much more with them than stick them into their ears.

That's a shame, though, because the clicker on the earbud cord has all kinds of magical powers.

- **When a call comes in:** You can click the clicker on the earbud cord to answer the call and, at the end of the chat, to hang up. (The earbuds have a microphone built in, too, so you can have the whole conversation without taking them off.)

  Or, to ignore an incoming call, squeeze the clicker for two seconds. You hear two low beeps, and the call goes to voice mail.

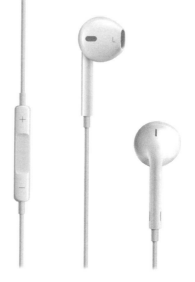

- **When you're listening to music:** Click the earbud clicker to pause the music or to start playing again. You can also pinch it *twice* to skip to the next song or *three* times to hear the previous song. (If you're watching a movie, the same gestures skip to the next or previous scene.)

  The + and – buttons on the clicker cord, as you'd guess, control the volume.

- **To speak to Siri:** Hold down the clicker on the earbud cord. Siri beeps twice as though to say, "OK, I'm listening!"

- **When you're taking a picture:** The volume buttons on the cord trigger the camera's shutter—a handy, impromptu remote control for selfies (see page 32).

- **When you're shooting a video:** You can start and stop recording by pressing the center clicker button.

The secret of the USB cable

USB cables are everywhere. They're those black cables that connect your computer to printers, scanners, phones, tablets, and cameras.

They're also a pain. There are so many possible different connectors on the end of them: the standard thin rectangular USB; the chunky square connector; the tiny flattened "Micro USB" type; and even proprietary shapes for certain gadgets.

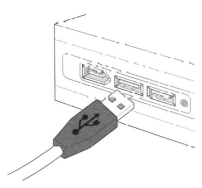

Worst of all, you can never tell which side is up! Half the time, you try to plug something into your computer—and you guess wrong. You can't plug it.

Those days are over now, because you're about to learn the secret. Before you plug, look down. Only one side has the forked USB logo on it, either painted or molded into the plastic. That's the *top* side.

How to save money on anything

Time equals money, of course; that's especially true when you buy stuff online. Just a *little* time can save you a *lot* of money.

Whenever you're about to buy something online, your first stop should be RetailMeNot.com. It's an Internet clearinghouse for discounts and deals from 50,000 stores online. Into the Search box, type what you're shopping for, and boom: You save an average of $20 per order.

Sometimes the site provides a promo code or coupon code; you then head to the retailer's Web site and paste that code into the Promo Code or Coupon Code box at checkout. Other times, you get a physical coupon that you print and take to an actual store.

Amazon.com offers enormous deals, too. The Amazon Warehouse offers pretty incredible "open box" specials (somebody bought a product, opened it, and returned it without using it) in every category: computers, cameras, phones, TVs, video games, shoes, sporting goods, and on and on. (You get there via *http://www.amazon.com/b?node=1267877011*—or by just typing *amazon warehouse* into Google.)

What to do when you've forgotten your charger

Electronic gadgets are great and all, but they have one big downside in common: They require *charging*. Everywhere you go with your laptop, phone, or tablet, you also have to carry its charger. Inevitably, you wind up in some hotel without the one you need. You forgot to pack it. You left it somewhere.

The great thing is that *you're not the first person to leave a charger behind*. Lots of people have left *theirs* behind—in the exact same hotel where you are!

So hie thee down to the front desk and ask. They'll offer you a Lost and Found box of iPhone chargers, Micro USB cables (for Android phones, e-book readers, some cameras), and even laptop adapters of every description. Borrow the one you need.

Meanwhile, take this moment to write your *own* name and phone number on *your* chargers, so you have a chance of recovering it when you lose it.

Part 2

The Computer

Chapter 5: Computers

Tablets and cell phones may outnumber laptops and PCs under the Christmas tree every season, but Macs and PCs have been around far longer. So they're far more mature—and have been saddled with far more features over the years.

That's too bad for anyone who likes life simple.

But it's good news for people who write books about how to figure it all out.

Why you should never turn off your laptop

Every day, several *times* a day, hundreds of thousands of people finish using their laptops—by *shutting them down*. And they shouldn't.

As it turns out, life is short. Fully shutting down a laptop means *waiting*—for the files and windows and programs to close. When you turn it on next time, you wait again—for the laptop to start up. And then you have to reopen the files and windows and programs you were working on.

Instead, each time a work session is over, just *close the lid*.

Closing the lid puts the laptop into Sleep mode. It's quiet; it's still; it's dark. It's using only a tiny trickle of power—enough to last for days. But when you open the lid to wake the laptop the next time, you're *instantly* right back to what you were doing. No quitting, waiting, reopening, waiting.

So if closing the lid is the best way to stop working, why does every laptop even *have* commands called Shut Down and Restart?

Because you *should* shut down all the way if you won't be using the laptop for a few days. And you *should* restart the computer from time to time, because the start-up process involves various system checks.

But for during the week, your "My work is done here" ritual should involve just closing the lid.

If in doubt, right-click

If you already know about right-clicking, congratulations! You're a power user. Skip ahead.

But Microsoft's research shows that a huge percentage of computer users don't know anything about right-clicking. So here's a refresher.

On both Mac and Windows, *thousands* of useful functions are hidden in shortcut menus—menus that pop out of something on the screen, like this one from the Recycle Bin:

It's important to know about shortcut menus. Often, the thing you want to do is available *only* in a shortcut menu.

So here's the drill:

- **Windows:** If you have a mouse, click the *right* mouse button. If you have a laptop, the trackpad probably has two areas on it—one for regular-clicking and the other for right-clicking. Or maybe it has a dedicated clicky button for right-clicking. Different laptops offer different methods (see page 155).

- **Mac:** If you have a trackpad, you can right-click by clicking with *two fingers*. (There are other ways, too, but that's the simplest.)

If you have an Apple mouse, it may appear to have only one button—but it actually has a *secret* right-click mouse button. It doesn't work unless you ask for it.

To do that, open the menu and choose System Preferences. Click Mouse. There, in all their splendor, are the options that control your mouse.

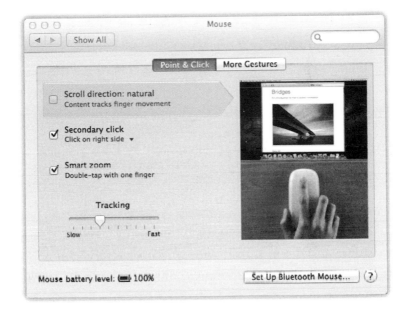

Turn on "Secondary click." (It's not called "right click" because left-handers might prefer to reverse the right and left functions. If you're a leftie, you can use the pop-up menu to specify that the left button is the "right-click.")

From now on, even though there aren't two visible mouse buttons, your mouse does, in fact, register a left-click or a right-click depending on which side of the mouse you push down.

On any kind of Mac, here's a backup plan: You can also right-click by holding down the Control key—bottom row—as you click the mouse on your target.

Now you're set. Next time you want to delete a photo, get a synonym for a selected word, create a new folder, or manipulate something on the screen in some other way, remember: Right-clicking is probably the answer.

The universal "Oops" key

You might already realize that there's an Undo command in almost every program. Use it when you've pasted the wrong thing, when you've deleted something accidentally, when you've changed your mind about renaming something.

Learn, then, if you haven't already, the Undo *keystroke*, which is Control+Z (on Windows) or ⌘-Z (on the Mac). It takes back the last thing you did. It often works even when you can't imagine that it would—say, when you've put something into the Trash or Recycle Bin or you've closed a Web browser window that you now want back.

Learn it. Use it. In time, it will become so reflexive that you'll find yourself using it in everyday life—even when you knock over a cup of coffee.

(No grief about how obvious and universally known this point is, either. There *are* people who don't know about the Undo keystroke, and their lives have just been changed.)

The universal "Yes" and "No" keys

Remember: The Golden Rule for Maximum Productivity says, "Keep your hands on the keyboard." Every time you reach for the mouse, you interrupt your flow of genius.

That's why you should learn the "Yes" and "No" keys on your keyboard, otherwise known as the Enter key (right side) and Esc key (top left).

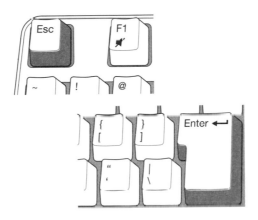

Pressing Enter is the same thing as clicking the most prominent button in any message on the screen, like OK, Save, Print, or Search. You always know which button that is, because it has a special border or color, like the ones shown at the top of the next page:

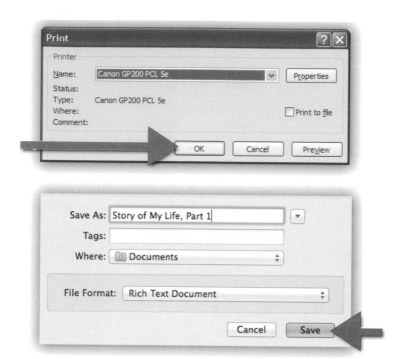

The Esc key is short for "Escape"—and it says *no* to any message or dialog box. It means "Close this" or "Cancel this." In the dialog boxes shown above, pressing the Esc key would "click" the Cancel buttons. This key also closes a menu you've opened (including the Start menu in Windows). And it makes a full-screen YouTube video shrink back down to regular size.

Once you've Entered the world of keyboard efficiency, there's no Esc.

Don't drag the mouse across text to select it

When you're trying to highlight some text—so that you can make it bold, or copy it, or delete it—don't bother dragging your mouse sideways across it. That's a frustrating, imprecise, slow way of going about it.

Unless you're paid by the hour, you should use these tricks instead:

- **Select one word by double-clicking it.** No matter which program you're using—even if you're on a phone or tablet—that one word is neatly highlighted.

Dear Taylor:

You haven't returned my 37 phone calls, and I'm feeling some panic.

Dear Taylor:

You haven't returned my 37 phone calls, and I'm feeling some panic.

- **Select more than one word by double-clicking the first one.** Then, with the mouse button still down on the second click, drag sideways. You select the text in one-word chunks.

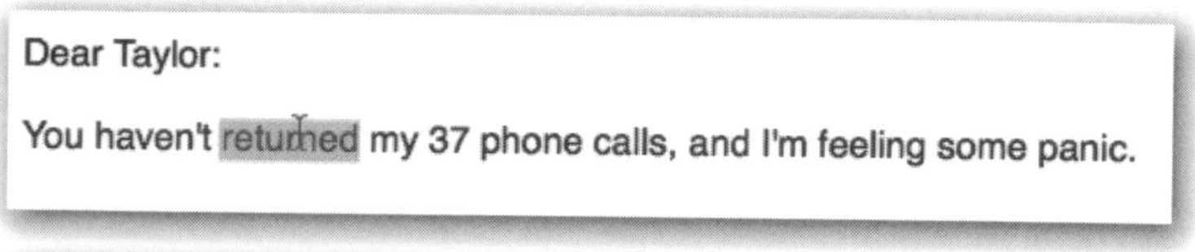

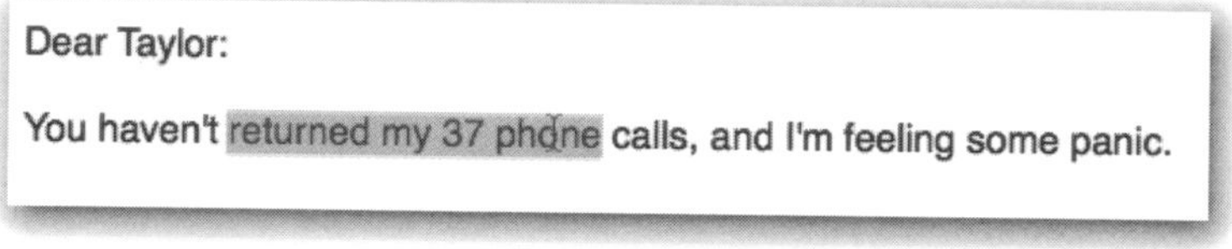

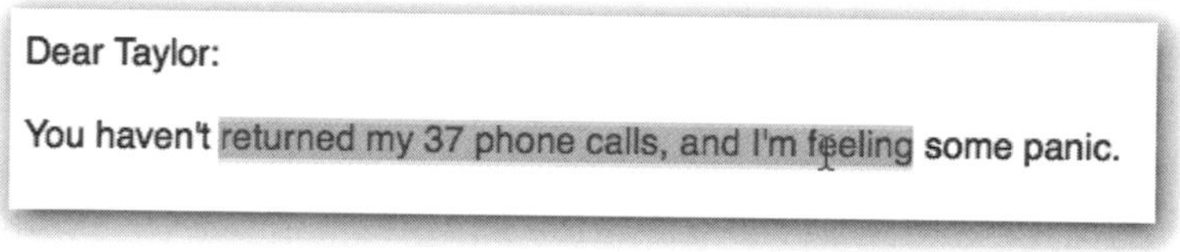

Dear Taylor:

You haven't returned my 37 phone calls, and I'm feeling some panic.

Dear Taylor:

You haven't returned my 37 phone calls, and I'm feeling some panic.

Dear Taylor:

You haven't returned my 37 phone calls, and I'm feeling some panic.

- **Triple-click to select one paragraph.** Quick and tidy.

Don't delete selected text before typing something new

Whenever you highlight some text (in an e-mail, an outgoing text message, or a word processor, for example), *you can type right over it.*

You don't have to delete it first.

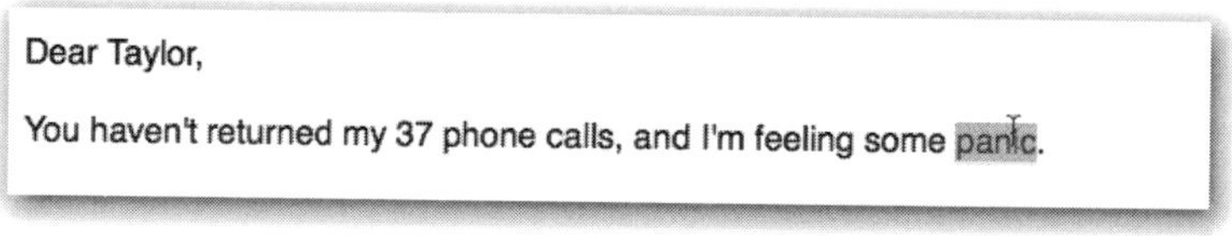

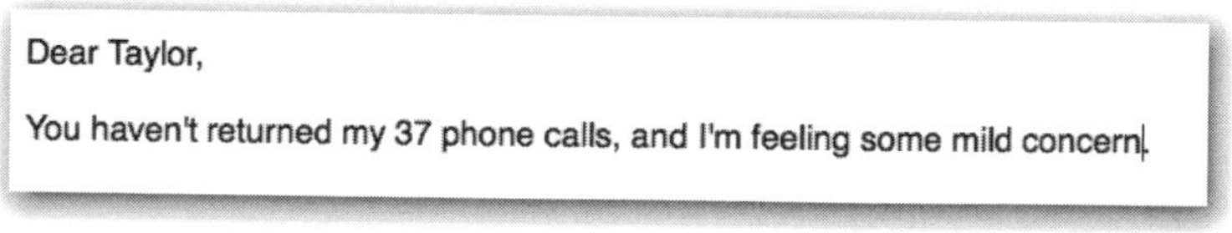

Dear Taylor,

You haven't returned my 37 phone calls, and I'm feeling some panic.

Dear Taylor,

You haven't returned my 37 phone calls, and I'm feeling some mild concern.

This trick works especially well in your Web browser. You click in the address bar so that you can type in the new address, right? And the address changes color to indicate that you've highlighted it.

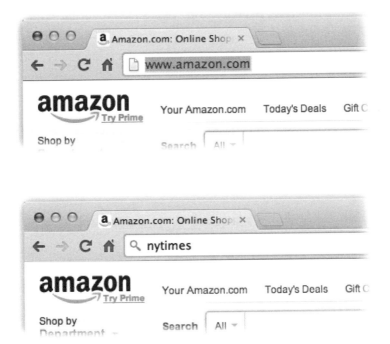

Your instinct might now lead you to press the Delete or Backspace keys—but that's a waste of calories. *Just start typing.* Whether you're on a computer, phone, or tablet, your machine knows that you mean to *replace* the highlighted text.

Nothing's gone until you take out the trash

Every year, certain people run out of space on their hard drives, despite having practically no files. It's because over the years, they've put 79 gigabytes' worth of stuff in the Recycle Bin or the Trash and never emptied it.

You probably know that you get rid of a file or a folder by dragging it onto the Trash icon or the Recycle Bin icon.

But that step doesn't actually delete anything. You still have to *empty* the Trash or Recycle Bin.

To do that, right-click the Trash icon (or the Recycle Bin), as described on page 78. From the shortcut menu, choose Empty Trash (or Empty Recycle Bin). Now that stuff is *really* gone.

You can resurrect deleted files

Even after you empty the Trash or the Recycle Bin, the files you put there aren't *really* really gone, no matter what you read on the previous page. Technically, your computer has only marked those files' spaces on your hard drive as "available to store new files." It hasn't actually erased them.

That's good to know if you ever delete an important file, you have no backup, and you're hysterical.

The first thing to do is *stop*. Don't do any more work on your computer. If you do, you might save some files onto the hard drive in the exact spot where your freshly deleted files once sat—and at that point, the deleted files have gone to the great CompUSA in the sky.

Before then, however, you can still retrieve them. Plenty of special recovery programs, like Stellar Phoenix (for Mac and Windows), can usually bring back the dead.

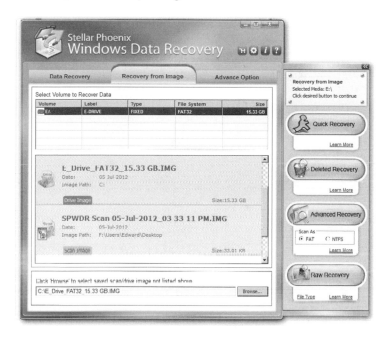

Even if they fail, you still have options. You can send your computer or hard drive away to an emergency file-resurrection company like DriveSavers, where clean-room technicians can take your drive apart and use rocket-science-y techniques to recover the lost files. Those services cost hundreds or thousands of dollars, but they have a stunning record of file resurrection.

How to really delete files if you work for the CIA

The fact that deleting files doesn't *really* delete them is great, because it leaves you a safety net if disaster strikes.

But it's not so great if the files you're trying to delete are extremely sensitive, confidential, or embarrassing. Those are files you don't want *anyone*, not even professionals, to revive.

On the Mac, that's no problem. From the Finder menu, choose the Secure Empty Trash command. The Mac doesn't just obliterate the parking spaces around the dead file. It actually records *new* information over the old—random *0*s and *1*s. Pure static gibberish. It re-scrubs that parking space *seven times,* actually. Whatever was in the Trash is now deleted irrevocably, irretrievably, forever.

On a Windows PC, you can download a free program like Eraser. It does the same thing: deletes the files you never want anyone to see again *and* saves nonsense data on the spot where they once were.

Type-selecting to find a file

When you're confronted by a crowded open folder on your Mac or PC, you can pluck a file out of a haystack by typing the first couple letters of its name.

Suppose that in one of your folders, there's a photo called Twilight. *Somewhere.* You can't even see it without scrolling.

But if you type *twi,* or maybe just *tw,* your computer finds that file, highlights it, and drops it at your feet.

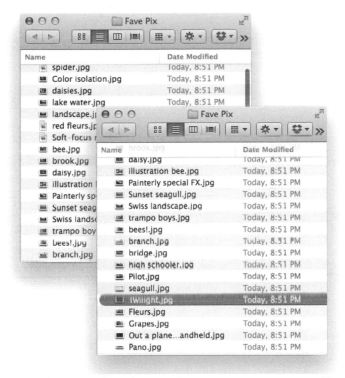

Same thing works when you're saving or opening a file. In this list box, you can type-select to jump to a folder. You can type *pl* to highlight the Places folder, for example—and then press Enter to open it.

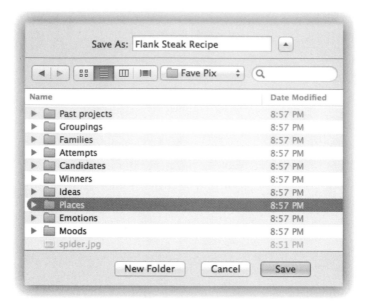

In Windows, a second possibility awaits: You can type the same letter over and over to jump from one match to the next. For example, if you have a folder that contains documents called Cactus, Comedy, and Cuticles, you can press the C key over and over to highlight them in succession.

Select all, select some, select none

It could happen to you. There's a list of files. You want to move, copy, or delete some of them but not all. But before you can move, copy, or delete them, you have to *select* them—highlight them. Obviously, you can click to select *one*. But how about two? How about all *but* two? Surely there's a faster way.

Of course there is.

- **Select a few consecutive items:** If you're looking at the contents of a window as a list, you can select several of them in a row. Click the first icon. Then, while pressing Shift, click the last one. You've just selected those files and all the files in between your clicks.

Click . . .

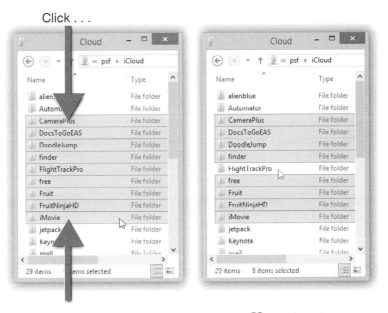

then Shift-click   Ctrl+click (⌘-click) to deselect . . .

- **Select a few nonconsecutive items:** If you want to highlight only the first, third, and seventh icons in a window, for example, start by clicking icon No. 1. Then hold down the ⌘ key (Mac) or the Ctrl key (Windows) as you click each of the others. Each icon darkens to show that you've selected it.

You can also ⌘-click or Ctrl-click a selected icon to *un*select it at this point, which is great if you've added one by mistake.

- **Select everything:** To select all the icons in a window, press ⌘-A (Mac) or Ctrl+A (Windows). That's the shortcut for the Edit→Select All command.

- **Select all but a few items:** Press ⌘-A or Ctrl+A to highlight all of the icons in the window. Now ⌘-click any unwanted icons to deselect them.

Shift-clicking words or numbers

The same principle of Shift-clicking also applies when you're trying to select *text*—in a word processor, e-mail, a Web page, or whatever. It's an especially important technique if you're trying to select a *lot* of text, when it's impractical to keep your mouse button down as you scroll a long way.

So here's how it works:

Double-click the first word you want to select. (You may recall that double-clicking a word selects that entire word neatly.) Or just click once in front of it.

Now get to the *last* word. That may mean scrolling. In some programs, it might even mean turning the page.

Finally, press Shift as you click the *last* word you want to select. And presto: Everything in between is now selected. (This same trick works in spreadsheets like Microsoft Excel.)

Enjoy that little time-saving stunt, by the way. You can't do it on a phone or tablet.

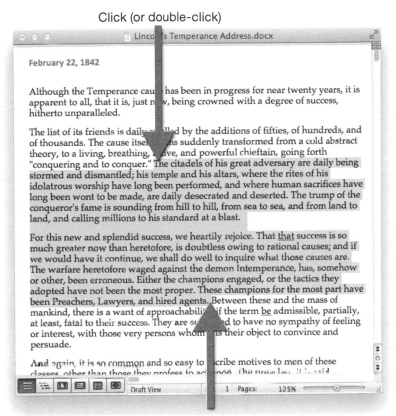

Click (or double-click)

Shift-click

Where your lost file went

Thousands of times a year, people create new documents, save them, name them—and then never see them again. They—the files, not the people—fall into deeply nested folders somewhere.

The problem arises at the moment you choose the Save command from the File menu. At this point, the computer shows you precisely where it proposes to put your newly created document. On both Mac and Windows, that folder is *usually* the Documents folder—a folder that was created expressly to solve the "Where'd my file go?" problem.

(If you have Windows 8 or later, all of this is more complicated; see the tip on page 96.)

For many people, keeping everything in Documents is a terrific suggestion. It means that you'll always know where to look.

And how do you look in Documents?

- **Mac**: Documents is one of the icons in the Sidebar, the list of folders at the left side of any desktop window. Click Documents once to see what's in it—and retrieve your lost masterpieces.

- **Windows:** Press ⊞+E to open a desktop window. Double-click the Documents folder.

If it's too late for all that and your file is already lost—well, you can always use your Find command

- **Mac:** Click the 🔍 in the top right corner of your screen (or press ⌘-spacebar) to open the Search box. Type a few letters of what you're looking for, either its name or words that you know are *inside* that file.

- **Windows 8:** Press the ⊞ key to open TileWorld (page 168). Type a few letters of the file's name or contents, and then press Enter.

- **Windows 7:** Click the Start menu (lower-left corner). Choose Search. Type a few letters of the file's name or contents, and then press Enter.

In each case, you see a list of search results—one of which is the file you're missing.
Usually.

The OneDrive (SkyDrive)

Whether you're aware of it or not, if you have Windows 8 or later, you're the proud owner of a *OneDrive*.

Before 2014, it was called the SkyDrive, which was a much more descriptive name. It's like a hard drive in the sky—or, less poetically, on the Internet.

The OneDrive is a reliable "disk" for storing important files—and convenient, since you can open it using any kind of computer, tablet, or phone, wherever you happen to be. (You just need the OneDrive app for your kind of machine.) The OneDrive also makes a handy parking spot for files you're trying to move from one computer to another—even files that are much too big to send by e-mail.

In Windows 8.1 and later, the OneDrive is built right in. Any time you save a file, Windows encourages you to save it into the Documents folder on your *OneDrive*. It's actually a good idea, as long as there's room on your OneDrive. (You get 15 gigabytes of storage free; you can pay to get more.)

Then again, if you spend most of your time offline, or if you don't particularly trust Microsoft, you may prefer to store your new files right on your computer instead of on your OneDrive. In that case, you can turn off that OneDrive/SkyDrive business.

To do that, move your mouse to the top right of the screen, so that the Charms panel opens. Click Settings, then Change PC settings, then SkyDrive, then turn off "Save documents to SkyDrive by default." That's the end of *that*.

The scroll bar class you missed

Sometimes, a document is so short, you can read it in a single window. ("Noble Acts of Congress in the Twenty-First Century" comes to mind.)

The rest of the time, the text is too tall to fit your screen. You have to *scroll* to move through it.

If you're an efficiency aficionado, you do that by pressing keys on the keyboard. You press the PageUp and PageDown keys, for example.

But you'll gain more control if you learn to use the *scroll bar* at the right side of any long window, not to mention the satisfaction of knowing what the hell you're doing.

The scroll bar is a miniature map of the entire document. In the middle, there's a sliding dark rectangle or blob. Its size represents how much of the page you're already seeing. For example, if the handle is one-quarter the height of the whole window, then one-quarter of the page is now visible.

You can operate the scroll bar in three ways. First, you can drag the handle up or down to move through the document. Second, you can click the tiny up or down arrows to scroll by one *line* at a time. Finally, you can *click in the track* above or below the handle; each time, you scroll by one screenful.

(You may sometimes see *horizontal* scroll bars, too. The instructions above apply to them equally well; just turn your head ninety degrees.)

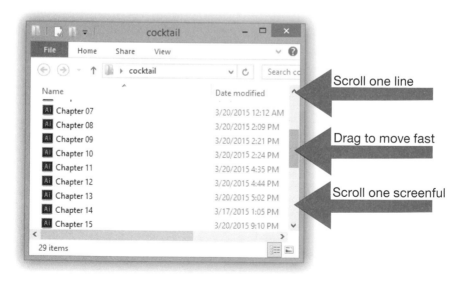

The alphabetical order of numbers

Computers are supergood at sorting things. In any window full of files, you can sort files alphabetically, by date, by size, and so on. (Just click the heading you want as the sort order. A tiny triangle points up or down, helping you remember in which direction the sorting goes.)

It's handy to know that, in alphabetical order, *numbers* come before letters. You can use that trick to force files or folders into any order you want.

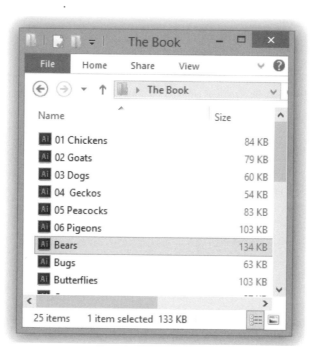

And how do you rename a file or folder? Click once on its name, then whip the mouse away. Or, on Windows, click once on the icon, then press F2. Either way, you can now type a new name.

Flip through open programs from the keyboard

How do you switch among open programs? If you feel that life is already too short, you should know the keystroke for that: The important one is the Alt key on Windows, or the ⌘ key on Macs.

While pressing the Alt ⌘ key, press Tab. A floating palette appears, bearing the icons of all running programs. Each time you press Tab again—still pressing Alt ⌘—you highlight the next icon. When you release the keys, the highlighted program jumps to the front.

Mac

Windows

If you just press Alt+Tab or ⌘-Tab briefly (instead of holding down anything), you don't see the row of program icons; instead, you flip back to the most recent program you had opened. That's a great way of hopping between two open programs—copying from a Web browser into an e-mail you're writing, for example.

Flip through open documents from the keyboard

If you read the previous tip (and maybe even if you didn't), you now know about the Alt ⌘-Tab trick. It lets you flip among open *programs*—from Word to Excel to your e-mail, for example.

But there's a far lesser known trick that lets you flip among open documents in the *same* program. Maybe you have six Web browser windows open, or five chapters of your novel in Word, or three photos in Photoshop. Here's the keystroke that lets you jump among them:

- **Mac:** While pressing ⌘, tap the ~ key (top row, next to the 1 key).

- **Windows:** While pressing the Ctrl key, tap the Tab key. (Works in *most* programs.)

Handy city!

Why there's a wheel on your mouse

The typical Windows computer mouse has two clicker buttons—and a little turning wheel in between them.

It's there for scrolling through a window, but it can also perform all kinds of other tricks. For example:

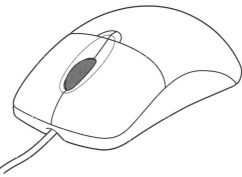

- **Magnify or shrink the page:** Press the Ctrl key as you turn the wheel.

- **Zoom forward or backward through a multipage document:** Press the Shift key as you turn the wheel.

- **Middle-click:** Click down on the wheel like a button. Yes, some mice actually have three buttons. A middle click is especially useful in Web browsers. Middle-click a link to open it in a new tab; middle-click an open tab to close it.

- **Enter turbo navigation mode:** Press down on the wheel and keep it down. A unique double-arrow icon appears on the screen; at this point, you can whip through a long document by moving the mouse forward or back. The farther you move from the icon, the faster you scroll.

- -

The Space bar = Play/Stop

What's the first thing you'd want to do after opening a music file or a movie file? *Play* it, of course.

That's why, on both Mac and Windows, the Space bar plays the special role of Play/Pause button. It works in any program that can play music or video, like QuickTime Player, Final Cut, iMovie, Windows Media Player, iTunes, and so on. It also works on videos at YouTube, Vimeo, and other video Web sites—usually. See page 286.

Free, instant tech support on Google

This one might *seem* so obvious that it's not worth the ink to print it. But thousands of people are unaware of it: Google is the world's greatest tech-support department.

Don't waste your time calling a company's 800 number. Don't bother looking on its Support pages. Don't pay some techie to help you. *Look it up on Google first!*

Whatever your computer problem, somebody else has had it before. And you can find the solution with Google. Every single time!

Here are some examples of what you can type:

- **page numbers won't print in Microsoft Word**
- **can't turn off gridlines in Photoshop**
- **how do I change ink cartridge in Canon Pixma iP7220**
- **Apple TV can't connect to iPad**
- **how do I delete photos from galaxy s4 phone**

Bonus tip: Add "solved" to your query, like this: "ipad won't charge solved." That way, Google will show you only the discussions where the question actually wound up answered.

Free, instant tech support on YouTube

When you're having trouble using a *feature* of your technology, try your query on YouTube. Often, you get to watch a *video* of someone else using that feature, and you'll realize what you've been doing wrong.

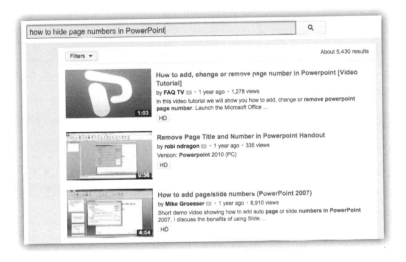

The magic of Dropbox

Dropbox is a free service that's so fantastic, it should be part of everyone's life.

To get it, go to Dropbox.com and click Install. You wind up with a magic folder on your computer desktop. Anything you put into this folder appears, within seconds, in an identical folder on each of your other computers.

And phones, and tablets. A file in your Dropbox is, in essence, in many places at once.

Phone

Computer

You can work on a project at home, then go to the office and pick right up from where you left off. You never have to send, carry, or transfer your files. They're waiting for you on every machine you use.

And even if you have only one machine, Dropbox makes a great backup system. If anything happens to your computer, all your files are safe and waiting on Dropbox.com.

But wait, there's more. You can share a Dropbox folder with other people—even if there are huge files inside. They can see, use, and edit your files, with permission, but without having to worry about sending or receiving them.

The free service gives you 2 gigabytes of storage. You can pay a monthly fee for far greater storage.

What the three dots mean in menu

The commands on your computer usually sit in menus, right? Those words at the top of the screen or the window—File, Edit, View, and so on.

If you click one of those menu names, you'll notice that some commands have three dots (...) after them, and some don't. (Typography geeks refer to that three-dot thing as an *ellipsis*.)

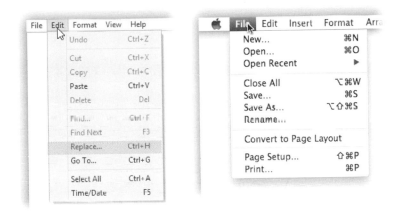

As it turns out, there's a reason for that. A command with an ellipsis means that when you choose it, a dialog box will open before any action is carried out. The ellipsis tells you, "You can't mess up by choosing me, because all I'm going to do is ask you further questions." It's an invitation to explore—and sometimes to discover—cool new functions you didn't know you had.

Teach yourself the keyboard shortcuts

O ne of this book's greatest missions is to persuade you to keep your hands on the keyboard—that learning a few choice keyboard shortcuts can save a lot of time and fiddling.

But the truth is, you're perfectly capable of learning a few on your own. Every program you use has a built-in *cheat sheet* of its own keyboard shortcuts. A cheat sheet called ... the *menus*.

Inspect these menus here, for example:

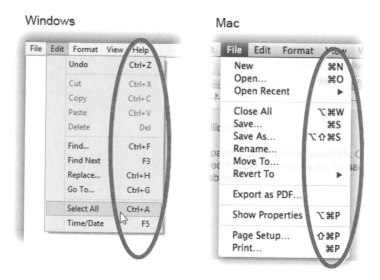

And there they are, listed down the right side: the keystroke equivalents of each menu command.

In Windows, those little notations are pretty clear. Alt means the Alt key, Ctrl means the Ctrl key. So the shortcut for Select All (in the example above at left) means "While pressing the Ctrl key, type A."

On the Mac, the symbols that appear in the menus can be a little more cryptic. Here's what they mean:

- ⌘ No mystery when you see this symbol in a menu; the same symbol appears on the actual Mac keyboard. It's called the Command key, and there's one on each side of the Space bar.

- ⇧ stands for the Shift key.

- ⌥ refers to the Option key (also labeled Alt on some Mac keyboards). You guessed that, didn't you?

- ⇪ means the Caps Lock key.

- ⌃ denotes the Control key.

- ⇞ and ⇟ refer to the PageUp and PageDown keys.

Now that you have this cheat sheet, you can interpret the illustration above at right. It tells you that the keyboard shortcut for Close All is ⌥-⌘-W. In other words, while pressing Option and ⌘, press the W key.

- -

Force-quitting a stubborn program

Ordinarily, when a program starts acting up, you can fix things by quitting the program and reopening it. But what happens if the app is so botched up that you can't even get to the Quit command? What if it's locked up so badly that the mouse and keyboard have no effect on it?

In those situations, you want to *force quit* that program, which means "jettison it from the computer with brute force."

- **Mac:** Press Option-⌘-Esc key. The Force Quit dialog box appears, listing all open programs. Click the name of the frozen one, and then click Force Quit.

- **Windows 8 and later:** Press Ctrl+Shift+Esc. The Task Manager dialog box opens. Click the stuck program's name and then hit "End task" to close it.

- **Windows 7 and earlier:** Press Ctrl+Alt+Del. The Task Manager dialog box opens. Click the stuck program's name; click the "End task" button. Click the End Now button if it appears.

In each case, the stuck program exits immediately, no matter how frozen it was. You can cheerfully reopen it and give it another shot.

Chapter 6: The Mac

Microsoft makes the Windows operating system, but other companies make the actual computers.

That's why Apple has an unfair advantage: It makes both the hardware and the software. It can design them to work well together, to work consistently, to work easily. It's that polish and elegance that explains why Mac owners tend to be an especially passionate bunch. People don't just call them "Mac users"; they get special names like fanboys, Macolytes, and Macheads.

Which is usually fine with the Macheads.

Chapter 5 teems with juicy tidbits that are equally useful no matter which kind of computer you use.

But Macs teem with secrets all their own.

(A note: Many of them are in the Mac's settings program, called System Preferences. To open it, click the at the top left of your screen; from the menu, choose System Preferences. Or, if your Dock is visible—the row of icons across the bottom of your screen—you can generally open System Preferences by clicking its icon, which looks like silvery gears.)

How to skip the password screen

S hould you password-protect your Mac?
If it's a laptop, yes. It might get stolen, and people might root through your stuff. If it's a desktop Mac, maybe—for example, if you'd rather not have your family members or co-workers snooping.

But what if it's your kitchen computer, and there's nothing especially private on it? Or what if it's just you and your spouse, and you have nothing to hide from each other? In those cases, there's no particular reason to password-protect your Mac. The only person inconvenienced by a password is you.

To set up *no* password, open System Preferences. Click your name, then Change Password. (Click the second Change Password button too, if one appears.) If you already have a password, you're asked to supply it now, in the Old Password box. Leave the "New password" box and the Verify box empty.

From now on, whenever you're asked for your password, just leave the Password box empty. You'll be able to log in faster.

Scrolling on the trackpad

These days, 75 percent of all Macs sold are *laptops*. So you're much more likely to be using a trackpad than a mouse. Therefore, a crash course in using your trackpad is in order.

- **You can scroll through a page** by dragging *two fingers* up or down your trackpad.

- **You can right-click** by clicking the trackpad with *two fingers.*

- **You can go Forward or Back between Web pages** by swiping *horizontally* with *two fingers.*

- **You can click without any sound.** That's handy when you're working in bed beside a light sleeper. To set this up, open System Preferences (page 111). Click Trackpad. Turn on "Tap to click."

From now on, you can trigger a click just by *touching* the trackpad rather than pushing down the clicking surface fully. It's totally silent.

How to fix the reversed scrolling

On a new Mac, dragging two fingers *down* the trackpad moves you *down* in a document. That's quite a switch from the way Macs have worked since their invention, in 1984. Until now, scrolling *down* has moved you *up* in a document. (Easier to see than to read.)

It is now, however, more like the iPhone or the iPad, and that consistency is what Apple was going for.

Which way do you want it to work? That's up to you. Open System Preferences (page 111). Click Trackpad. Click the Scroll & Zoom tab.

See the option Scroll Direction: Natural? That should be *on* if you like the newer arrangement but *off* if you prefer the older one.

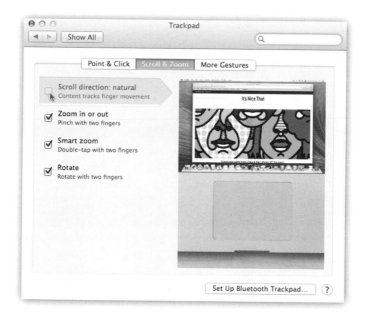

Make your windows stop reopening

In recent versions of the Mac operating system, you may have noticed a quirk: The documents that were open when you last quit a program magically *re*open, ready for you to get back to work. Everything is *exactly* as it was.

If you tend to work on the same documents day after day, all of this can be handy; the auto-reopened document helps you remember what you were in the middle of doing.

Then again, maybe this feature drives you crazy.

Fortunately, it's easy to turn it off: globally, on a per-program basis, or on a per-Quit basis.

- **Turn off all auto-reopening:** Open System Preferences (page 111). Click General. Turn on "Close windows when quitting an application."

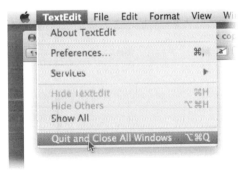

- **Turn off reopening next time:** If you hold down the Option key while quitting a program (for example, when you open the Safari menu to quit, the Quit command changes to Quit and Close All Windows. The next time you open the program, it won't remember your window setup.

(Unless you've turned *off* the "Restore windows" feature in System Preferences, as described above. In that case, pressing Option makes the Quit command read Quit and *Keep Windows*.)

* **Forget the window setup as you open a program:** If you forgot to use the tip in the previous paragraph, you can force a program to forget its previous window setup by pressing the Shift key as the program opens.

Make the Mac read anything out loud to you

Your Mac can read aloud to you. It's a great way to catch up on the news while you're making breakfast, or to proofread something you've written. You'll catch all kinds of mistakes when it's read to you out loud.

Many Mac programs have a Start Speaking command right there in the Edit menu. From the Edit menu, choose Speech, then Start Speaking. These programs include Mail, Safari, Messages, Stickies, Pages, and TextEdit.

But in fact, the Mac can read *anything* to you, in any window of *any* program.

To set this up, open System Preferences (page 111) and choose Dictation & Speech. Click the Text to Speech tab at the top of the window. Here's where you choose which voice you want the Mac to use—you get a choice of 18 different voice styles, ages 8 to 50—and how fast it should speak.

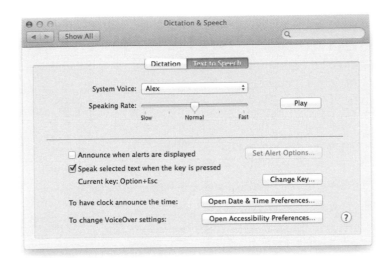

Your main job, though, is to turn on "Speak selected text when the key is pressed." By "the key," it means the keystroke Option-Esc (while pressing Option, tap the Esc key at top left).

(The Change Key button lets you replace that combo with a different one; just choose a keystroke that doesn't conflict with the program you're using. Try Control-T, for example.)

In any case, you can now close the System Preferences window.

Now you're ready for the magic. Find some text worth reading. Drag your cursor through some text to highlight it, and then press Option-Esc (or whichever keystroke you use). Instantly the Mac begins reading it aloud. To stop the playback, press the same keystroke again.

Pretty cool, huh?

Sign a contract electronically

These days, contracts and agreements are going electronic. Fortunately, your Mac can *take a picture* of your actual written signature. You can then zap it onto whatever PDF contract may come your way.

To teach it your signature, sign your name on a piece of white paper. Open the Preview program (it's in your Applications folder).

Open the Edit toolbar, which looks something like this:

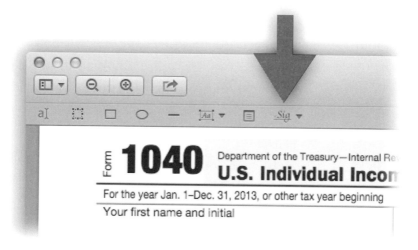

If you don't see it, open the View menu and choose Show Edit Toolbar (or Show Markup Toolbar).

Now click the Sig button (shown by the arrow above). If there's a pop-up button, choose Create Signature from Camera. (You'll see your actual camera name.) Hold up your signed paper to the Mac's camera. When the positioning looks good, click Accept or Done. You've just stored your signature.

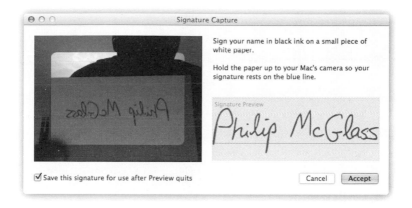

When the time comes to sign a document, make sure the Edit toolbar is visible. Open the pop-up menu shown by the arrow on the previous page, choose your signature's name, and then click on the document where you want to sign it. Boom— you've just signed the form.

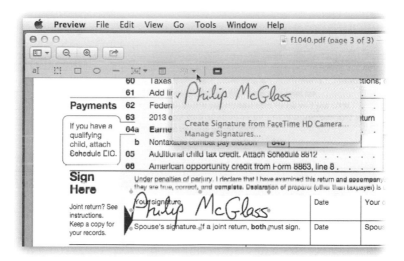

You can delete signatures (and add new ones), if you like. Open the Tools menu; choose Annotate, then Tools, then Manage Signatures. There are your + and – buttons.

See what's in a file without opening it (Quick Look)

The Mac's Quick Look feature is so useful and amazing, there ought to be a shrine to it.

Quick Look is a Finder desktop feature that lets you see what's in a document—see what's behind its icon—without having to open a program. Read the text inside it, play the movie, admire the photo, right there on the desktop.

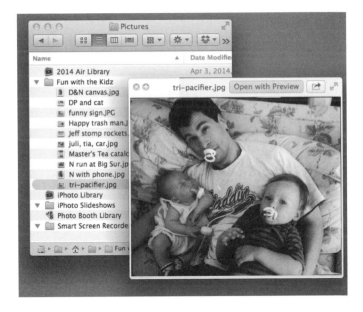

It's so important, in fact, that Apple has devoted the keyboard's biggest key to activating Quick Look: the Space bar. Click any document icon once and then press the Space bar to get a quick look, thanks to Quick Look. You get a huge, instantaneous preview of the document.

For example, you can read the fine text in a Word or Power-Point document without actually having to open Word or PowerPoint, which saves you about 45 minutes. If it's a movie or a sound, it plays instantly.

If it's a PDF, Word, PowerPoint, Excel, Keynote, or Pages document that's several pages long, you can actually turn the pages to read it. If you hold down the Option key, you can zoom in and out by swiping two fingers up or down your trackpad (or by scrolling the mouse wheel, ball, or scroll surface).

Once the Quick Look preview is open, you have two choices. You can press the Space bar *again* to close it—or you can click Open With [Program Name], the button in the top right. It might read, for example, "Open with Preview." And it means "Go ahead and open this file as though I'd double-clicked it in the first place." (And if you *click and hold* your cursor on that "Open with" button, you get a pop-up menu of other programs that could open this file.)

Type half as much

The Mac has an automatic abbreviation-expander feature. You can set it up so that when you type *addr*, the Mac types out:

Taylor DeLorenzo
12883 South New American Terrace
Southwest Flintrock, AZ 85003

Maybe you just set up one auto-expanding abbreviation; maybe you set up dozens. The point is that *whatever* you type regularly, your Mac can fill in for you.

To set this up, open System Preferences (page 111). Click Keyboard, then click Text. There's your list of ready-to-use abbreviations. Add a new one by clicking the + button below the list. Fill in the Replace and With columns with the abbreviation and the expanded version you want. Close the window.

From now on, that abbreviation expands automatically, anywhere you can type.

How to type accents and symbols

The Mac has always been a typographical powerhouse, and part of the reason is its ability to type all the symbols that don't appear on the keyboard: ©, ™, ¨, and so on.

The key to unlocking them is the Option key. Whenever you press Option, your ordinary, mild-mannered alphabet keys suddenly type fascinating, exotic symbols like these:

Option-g   ©

Option-r   ®

Option-2   TM

Option-4   ¢

Option-1   ¡ (Spanish exclamation)

Option-3   £ (pounds)

Option-7   ¶ (paragraph break)

Option-8   • (bullet)

Option-0   ° (degrees)

Option-p   π (pi)

Option-/   ÷ (division symbol)

The Option key gets you diacritical (accent) marks, too. The trick is to *hold down* the key for the letter you want to accent. For example, to type the last letter of *café*, hold down the letter *e* key until you see this palette pop out:

Click the symbol you want. (The same palette pops out of all five vowel keys, plus S, Z, C, and N.)

There are dozens more, though: math symbols, international symbols, fancy brackets, stars, and on and on. You'll never remember them all.

Fortunately, there's a cheat sheet. Here's how to set it up:

Open System Preferences (page 111). Click Keyboard. Click Input Sources. Turn on "Show Input menu in menu bar." Now look at the top right of your screen: You've installed a new mini-menu.

Next time you want to type a special symbol, open this menu and choose Show Character Viewer. The cheat sheet appears. (If it's full of smileys, click the tiny ⊞ icon at the top right.)

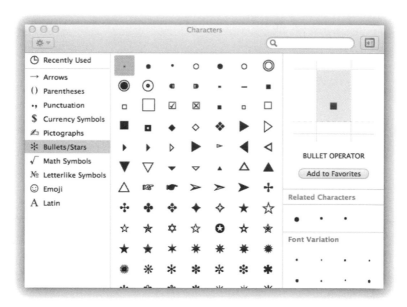

The symbols are arrayed before you in various category headings: Arrows, Bullets/Stars, Math Symbols, and so on.

Double-click a symbol to type it into your document.

Talk to type on your Mac

Not only can your Mac read text aloud to you—it can also write down what you say. You can speak to type.

The dictation feature is much faster than typing. And it's incapable of misspelling a word (although it may sometimes write the *wrong word entirely*).

To try it out, open any program where you can type, like your e-mail program or Word. Now tap the Fn key twice. (It's in the lower-left corner.)

The first time you try this, the Mac asks if you're sure you want to turn on dictation, and it offers to turn on Enhanced Dictation (a mode that doesn't require an Internet connection).

When you double-press the Fn key, a tiny microphone button appears next to the insertion point, and you hear a xylophone note. Now you can speak. Talk normally—don't be loud or slow. But do speak your punctuation, like this:

Dear John (comma) (new paragraph),

I've always loved you (dash)—well (comma), at least I've always liked you (period).

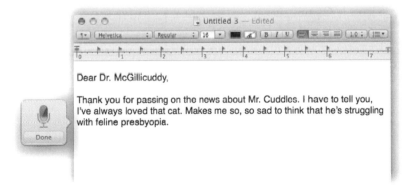

You can also speak symbols like "percent sign" (%), "at sign" (@), "dollar sign" ($), and so on. And you can say "cap" right before the word, like this: "Dear (cap) John, I forgot to mention that you're in my new book, (cap) The (cap) Loser."

After you finish speaking, click Done or press the Return key.

The Mac plays another xylophone note—higher, this time—and then the transcribed text appears, exactly as though you'd typed it. (By the way: You'll be far more accurate if you wear a headset microphone. You can buy one inexpensively from *emicrophones.com*, for example.)

The calculator staring you in the face

On your Mac, calculators are all around you. There's a Calculator program in your Applications folder. There's a mini-calculator on your Dashboard.

But when you need some quick math, the easiest-to-find calculator is the one that's on your screen all the time: the Q in the upper-right corner of your screen.

That, of course, is the Spotlight icon, the one you usually use for *searching* your Mac. But as it turns out, it's a dandy little calculator, too.

To open the Q feature, click it, or press ⌘-Space bar. Now type (or paste) the equation you want to solve. You might type, for example, 234+923-12*25. And there it is, the first item in the Spotlight results: 857.

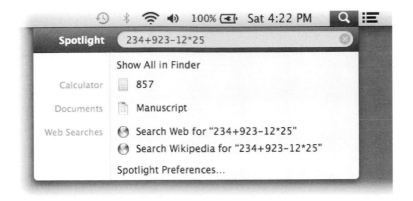

There's more to it than the four basic math functions. You can use square roots, for example: Type *sqrt(49)*, and you'll get the answer 7. It also works with the functions *log(x)*, *exp(x)*, *sin(x)*, *sinh(x)*, and *e*. You can even type *pi* when you need it (3.14159…).

Magnify the screen

There are all kinds of situations in which you might want to magnify the Mac's screen. Sometimes there's a video that you'd like to play bigger. Sometimes there's tiny type that you can't read. Sometimes you want to study how some icon or font character was designed, pixel by pixel. In any case, this fantastic feature lets you zoom *way* in—so much that a single word fills your entire screen.

You have to turn on this option in System Preferences (page 111). Click Accessibility, and then click Zoom. Finally, turn on "Use scroll gesture with modifier keys to zoom." Close this window.

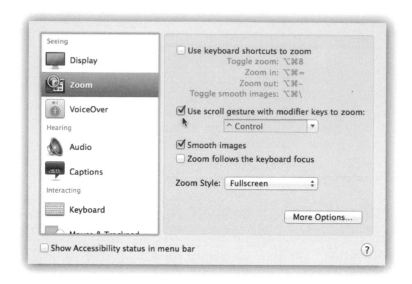

From now on, whenever you want to zoom into your screen, hold down the Control key—and drag two fingers up the trackpad. Or, if you have a mouse, hold down the Control key and turn the scroll wheel away from you. (If you have an Apple mouse, it may have a tiny scroll *ball* on top instead—or even a nonmoving scroll *area* on top. That's if you have the cordless, flattened-looking Magic Mouse.)

Control-drag downward (or Control-scroll downward) to zoom out again.

Take a picture of the Mac screen

No computer on earth offers more built-in methods for taking a *screenshot*—that is, capturing a printable picture of what's on the screen. Screenshots are great. You can take one when you want some tech-support person to see how messed up your software looks. Or when you see something online that you want to save forever. Or when you're writing up computer instructions for someone else, or maybe a review of some new software.

Here's how to capture:

- **The whole screen:** While holding down Shift and ⌘, press the number 3 key. You hear a *snap!* camera-shutter sound. And presto: There's a new graphics file on your desktop, called something like "Screen shot 2015-4-14 at 5.18.12 PM" (or whatever the date and time is).

  You can open this file, copy it, edit it, print it, e-mail it, and so on.

  Bonus tip: You can change that keystroke to something easier to remember. See the tip on page 140.

- **A rectangular area:** While holding down Shift and ⌘, press the number 4 key. Your pointer is now a + symbol, accompanied by tiny digital readouts—the coordinates of your cursor on the screen at that moment.

  Now drag diagonally across the screen, enclosing a rectangular area. When you let go, the camera-click sound plays and the screenshot file appears on your desktop—containing only the area you selected.

  Bonus tip: If you press Shift once you start dragging, the Mac limits your mouse movements to perfectly horizontal or perfectly vertical.

  Bonus tip 2: If you press Option once you start dragging, your selected rectangle grows from the *center* instead of the corner.

  Bonus tip 3: If you press the Space bar while you're dragging, the shape of your box freezes—but you can *move* it around the screen with your mouse.

Bonus tip 4: If you chicken out, tap the Esc key at any time during your drag. You've just canceled the whole business.

- **A neatly snipped window or open menu:** Once again, press Shift-⌘-4. When your pointer is a + symbol, press the Space bar. Your pointer now looks like a tiny camera!

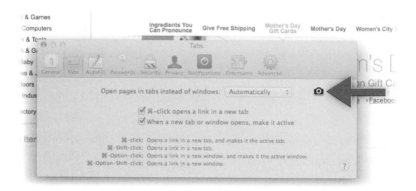

Move your cursor so that blue foggy highlighting fills the window or menu you want to capture, then click. The resulting picture file snips the window or menu neatly from its background.

Finally, a bonus tip: If you hold down the Control key as you click or drag using any of those methods, the Mac copies your screenshot to the invisible Clipboard, ready for pasting, instead of saving it as a new graphics file on your desktop.

Take a movie of the Mac screen

You can also record a *movie* of screen activity. It's a quick, easy way to "film" something for later: to demonstrate a technique, to preserve some glitch you want to complain about later, or even to save an online video (a YouTube video, for example) as a file to play back later.

The key to all of this is QuickTime Player, a program in your Applications folder. Open it. From the File menu, choose New Screen Recording. Click the red Record button.

A message appears now, letting you know that you can record either the entire screen (just click to begin) or only a *part* of the screen (drag diagonally to form a box around the part you want).

When you're finished recording, click the ■ button in the menu bar.

Click the Play button to make sure you've captured what you wanted—and then save it by opening the File menu and choosing Save. You wind up with a movie file that you can play any time you like, e-mail to someone, post to Facebook, or whatever.

How to copy files by dragging them

It's been confusing from the beginning: When you drag icons from one window to another on the Mac, are you *moving* them or making copies?

- **If they're going to a different disk,** you're *copying* them. (A green + symbol lets you know.)

  (Manual override: If you press the ⌘ key as you *release the mouse*, you *move* the files or folders instead, deleting them from the original disk.)

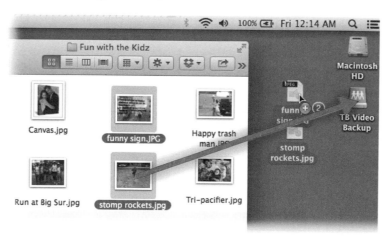

- If they're going from window to window on the same disk, you're moving them.

(Manual override: If you press the Option key as you *release the mouse*, you *copy* them instead of moving them—even if they're on the same disk or in the same window.)

In any of these cases, if you realize you've just made a mistake, press ⌘-Z (the shortcut for the Undo command) rewinds time to the moment just before your errant drag.

- -

Close all windows at once

One of the Mac's best features is its consistency. In every program, the same keystroke—⌘-W—closes any window.

But if you add the Option key, you close *all* windows in whichever program you're using. All your Web browser windows, all your desktop windows, all your e-mail windows.

The Option key often has that effect—making a window command apply to all open windows. For example, pressing

⌘-M usually *minimizes* a window (shrinks it into an icon on your Dock). But if you press Option-⌘-M, you minimize all windows simultaneously.

Toss an icon into the Trash without the mouse

You know how to throw away a file or a folder, right? You drag it onto the Trash icon on your Dock.

That, however, can get old quickly—if you have a big screen, for example, or if you're working through folders full of stuff and discarding one icon after another.

An amazing keystroke awaits: ⌘-Delete. Click the icon (or icons) you want to throw away, then press ⌘-Delete. They fly into the Trash all by themselves—no mousing required.

Send a file to a nearby Mac wirelessly

This is the *greatest*.

When you want to send a file or folder to someone else's Mac, you don't have to fiddle with e-mail, or networking, or setup. Thanks to a feature called AirDrop, your Mac can "see" any other Macs, iPhones, and iPads within 30 feet—and you can toss files their way, wirelessly, without any passwords or fuss. No Wi-Fi network is required; it works on the beach, on an airplane, or in a tent in the mountains.

(The Macs have to be running Mac OS X Lion, the 2011 Mac software, or later. AirDropping between Macs and iPhones requires the latest software on each—Yosemite and iOS 8.)

To use AirDrop, open any Finder (desktop) window. Click the AirDrop icon at the left side. (Or open the Go menu and choose AirDrop, if that's easier.) Your lucky recipient must do that, too. Now you *both* have the AirDrop windows open. After a moment, it shows icons for both of you.

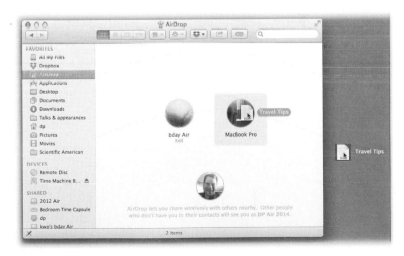

Now you can drag a file or folder icon (or a bunch of them) onto the other person's icon.

On both ends, the Mac asks if you want to proceed. You should click Send (or press Return); the recipient should click Save (or "Save and Open").

A progress bar curls around the other person's icon. The transfer is encrypted, so NSA spies can't intercept your files. Finally, the transferred file winds up in your buddy's Downloads folder.

Turn any document into a PDF file with one click

You've seen PDF files. They're great. They're incredibly common. People send around flyers, user guides, school papers, and all kinds of other documents in PDF format.

Why are they so handy? Because when you send other people a PDF file, they see exactly the same fonts, colors, and layout that you've created—even if their computers don't have the same fonts or software that you used.

PDF files are universal, too. Every kind of computer, phone, and tablet can open them. And you can search a PDF for certain words; it's not just a frozen picture.

Best of all, the Mac can turn *any* document into a PDF file. You don't have to buy a program to create PDF files, as Windows PC owners must.

And how do you do that? *Print* the document. That is, from the File menu, choose Print. In the dialog box that appears, inspect the lower-left corner: There's a PDF *menu button*. From it, choose Save as PDF—and the deed is done.

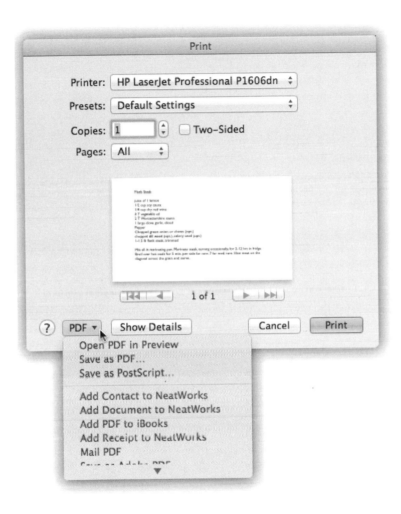

The best (and least known) Mac feature ever: Web receipts

Here's the unsung feature of the year: The "Save as Web Receipt" command.

Every time you buy something online, the transaction wraps up with a "Print this receipt for your records" screen. But don't. You'll waste paper and ink, and you won't be able to find the printout later.

Instead, open the File menu and choose Print. In the Print dialog box, click the PDF button and choose "Save PDF to Web Receipts Folder."

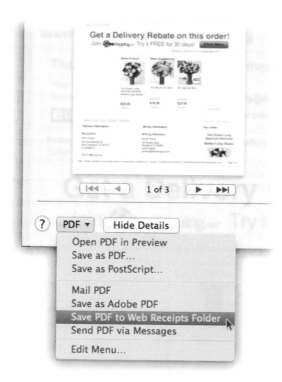

Instantly, the Mac creates a PDF version of your receipt and puts it into a special Web Receipts folder. (It's inside your Documents folder; see page 93.)

If you never need that receipt again, you can open the folder to read it or print it—or, better yet, you can *find* it by looking for words inside it. Just use the Mac's search feature (click the Q icon in the upper-right corner of your screen).

The universal fix-it step: Repair Disk Permissions

No computer works flawlessly all the time and forever. Too many cooks are working in that kitchen: the programmers who wrote the operating system, the ones that make the program you're using, and so on. All kinds of glitches, freezes, and visual oddities can occur.

You'd be surprised at how often you can clear up the problem using the Disk Utility program that came with your Mac. It's incredibly effective in fixing just these sorts of mysterious problems.

You'll find Disk Utility in your Applications folder, inside its Utilities folder. Next time you experience some mystifying glitch on your Mac, open that program. On the left side, click the name of your hard drive (it's probably Macintosh HD). Then click Repair Disk Permissions.

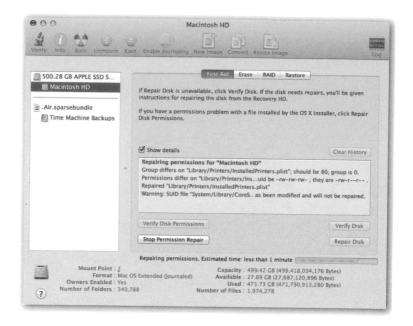

The fixing process takes a few minutes. But when it's all over, you'll probably see a list of things that Disk Utility has fixed—and you'll probably find that the mysterious problem has disappeared.

Change your keystrokes

If you're a keyboard-shortcut lover—and you should be!—it might come as a delightful surprise that you're allowed to *change* the key combination shortcut for any command in any program. Or, if there's some command that doesn't *have* a keyboard shortcut, you can add one.

At the desktop (in the Finder), for example, there's no keyboard shortcut for the Show Clipboard command. Maybe you think there should be!

Open System Preferences (page 111). Click Keyboard, then Shortcuts; click App Shortcuts in the left-side list.

Click the + button just beneath the list. From the Application pop-up menu, choose the program whose command you want to change. In this example, that'd be Finder. (If the program's name doesn't appear in the pop-up menu, then choose Other; navigate to, and double-click, the program you want.)

In the Menu Title box, type in the name of the menu command whose keyboard shortcut you want to change or add—*Show Clipboard,* for example. It must exactly match the spelling and capitalization as it appears in the menu, complete with the little three-dot ellipsis symbol (...) that may follow it. (You make that character by pressing Option-semicolon.)

Next, click in the Keyboard Shortcut box. Press the new or revised key combo you want. For Show Clipboard, you might choose Control-S.

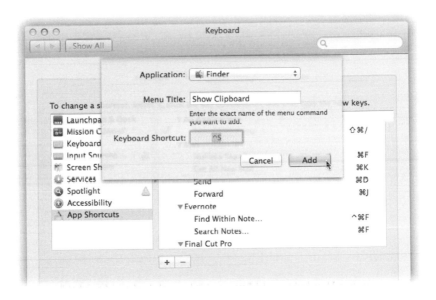

The Mac writes down your keystroke combo in the Keyboard Shortcut box, using the symbol notation described on page 109—unless, of course, the program already *uses* that combination for another command. In that case, you hear only an error beep; try again with a different combo.

Finally, click Add.

Scroll down in your Keyboard Shortcuts list; the keystroke you selected now appears under the appropriate program's flippy triangle. The next time you open the program you've edited, you'll see that the new keystroke is in place.

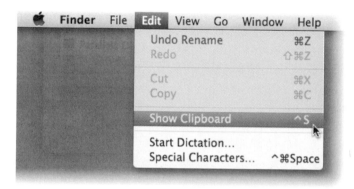

Let the Mac turn itself on and off each day and night

Your Mac can turn on and off automatically each day, according to your schedule. That arrangement conserves electricity, saves money, and reduces pollution. Since the Mac does all the starting up and shutting down automatically, you won't be inconvenienced in the least. You might actually never see it turn off again.

This option makes the most sense for desktop Macs; if you

have a laptop, you should finish each work session by just closing the lid. (The laptop goes into low-power Sleep mode, and it will wake instantly the next time you open the lid to work.)

To set things up, open System Preferences (page 111). Click Energy Saver. Click Schedule.

In the box that appears, turn on both checkboxes. Set up the pop-up menus and the times. For example, if you work from 8:30 a.m. to 5:30 p.m., you might want to set things up like this:

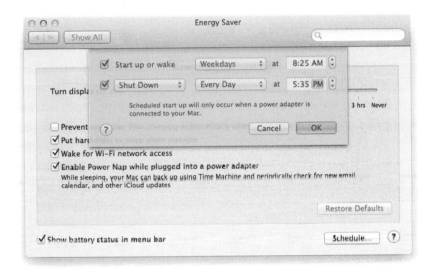

Understanding the F-keys at last

Each key on the top row of a Mac keyboard bears *two* painted labels. For example, the first three keys are labeled F1, F2, and F3—but also ☀, ☀, and ⊞.

That's Apple's way of telling you that each key has two personalities.

Out of the box, those keys control the Mac's hardware functions: screen brightness, keyboard brightness, speaker volume, and music playback. Press ☀ and ☼ to fiddle with the screen brightness, ◀ and ◀)) to adjust the speaker volume, and so on.

But some programs, especially Microsoft's, rely on those keys to trigger common commands. For example, F1 means Undo, F2 means cut, F3 means Copy, and F4 means Paste.

But how are you supposed to press F1 for Undo, when pressing that key dims the screen instead?

In those situations, you can just add the Fn key (lower left on Mac laptop keyboards, center block of keys on big Mac keyboards). It switches the function of the function keys; it lets F-keys be F-keys.

In other words, if you press Fn *and* F1, you get Undo, not the Dim Screen symbol (☀) function.

And what if you use those F-keys *more often* for Microsoft commands than you do for features like brightness and volume?

In that case, you can reverse the logic, so that the F keys *usually* operate as menu-command shortcuts, and require the Fn key only when you want the brightness/volume features.

To do that, open System Preferences. Click Keyboard. Turn on "Use all F1, F2, etc. as standard function keys."

And that's it. Now pressing the F-keys *alone* will trigger software functions. They control brightness and audio only when you're pressing Fn at the same time.

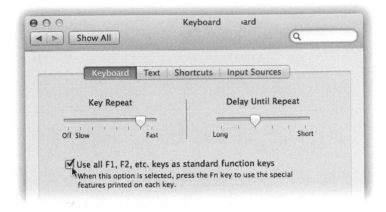

$100 for a year of private lessons

If you live near an Apple store, boy, does your computer company have a deal for you.

Apple's One to One program, which costs $100, gives you weekly, hour-long *private lessons* in the store with an Apple genius—for a whole year. That's $100 *for the year,* not per lesson.

If you want to get up to speed on your Mac or any of its programs, this, brothers and sisters, is the deal of the century.

Chapter 7: Windows

Ah, Windows. The most-used software in the history of mankind.

Windows has been around since 1985. Each new major version looks better, is more secure, takes advantage of better hardware—and adds more, more, more features. It's gotten to the point where you really need a book to know which ones are worth learning.

A few years ago, Microsoft observed the popularity of tablets like the iPad—and predicted that *all* computers would soon have touch screens.

So in 2012, Microsoft introduced Windows 8. It was two operating systems in one, superimposed. There was the regular Windows, the one whose desktop fills hundreds of millions of screens, with a software library of 4 million programs.

And then there was a new operating system for tablets, one that looks totally different, works differently, and requires all new apps.

Microsoft doesn't have a name for this mode (it abandoned the names Metro and Modern), so let's call it TileWorld.

It's an interesting experiment, but it means that you now have two Web browsers to learn, two different Help systems, two control panels, and so on.

On the following pages, you'll find tips and tricks to surviving both recent Windows versions: Windows 7 and its successors, Windows 8 and 8.1.

- -

The difference between Backspace and Del

Your keyboard probably has two keys that sound an awful lot alike: one called Backspace and one labeled Del or Delete.

They're actually different. Backspace is the one you're used to; it deletes the typed character to the *left* of the blinking insertion-point cursor. But pressing Del removes the character to its *right*, which can also be handy when you're editing.

That's why the Del key is sometimes more clearly referred to as the Forward Delete key—but that's too much writing to fit on the top of the key.

- -

Free, excellent antivirus software

If you have Windows, you need an antivirus program. Period. Most of these programs cost money. A lot. And you have to pay again every year.

There are some free ones, though—and one of them comes from Microsoft itself. It's easy to use, attractive, and it fights both viruses and spyware.

- **For Windows 7 and earlier:** The program is called Microsoft Security Essentials, and you have to download it yourself. Here's the address: *http://windows.microsoft.com/en-us/windows/security-essentials-download.*

- **For Windows 8 and later:** It's called Windows Defender, and it's built right into Windows.

However, it comes *turned off.* If you want this program to protect your machine, uninstall whatever trial software may have been nagging you to buy it (Norton, McAfee, or whatever). Then turn Defender *on.*

To do that, search for and open the Windows Defender program (type its name into the Start menu or at the Start screen). The big red "Turn on" button is right there in front of you.

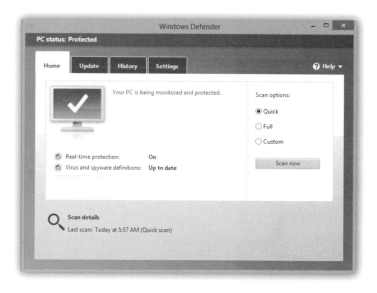

Security Essentials, or Defender, or whatever your version is called, continuously monitors your PC for infections. Microsoft sends it a daily update of its virus database, so that your anti-virus program will recognize new viruses and other bad stuff.

If the program finds something fishy, it usually deletes the virus or spyware automatically. Occasionally, it may ask your permission first.

There are other free antivirus programs, by the way, like Avast Free, AVG Free, and ZoneAlarm Free. They may nag you to buy the Pro versions, though, and there's nobody to call for help.

How to skip the password screen

When you turn on your PC, you're asked to log in—to provide the name and password for your account. That's an important feature, both for security and convenience (because it keeps *your* files and settings separate from everybody else's).

But what if there *is* no "everybody else"? What if you don't share the computer with anyone? Or if there's nothing private on it, or if you share it with someone you trust?

In that case, the name-and-password business is just a pointless roadblock—and you can eliminate it. You can *turn off* the requirement to log in with a password.

- **No password when waking the PC:** If you have Windows 8.1 or later, you can eliminate the requirement for entering a password when you *wake* the computer. (You still have to log in when you turn it on or restart it.)

  Setting it up requires a visit to the Accounts pane of PC Settings. The quickest way to get there is to type *account* at the Start screen. In the results list, select "Your account set-tings"; on the next screen, select "Sign-in options."

Finally, under "Password policy," select Change. In the warning box, tap Change again. That's it! From now on, you won't be asked for your password when you just wake the machine after it's gone to sleep.

- **No password required, ever:** If you're willing to do some technical fiddling, you can also set it up so that you *never* have to enter your password—even when you're starting up the machine.

Start by pressing ⊞+R. In the Run box, type *netplwiz*. Hit OK. Now you see the User Accounts dialog box:

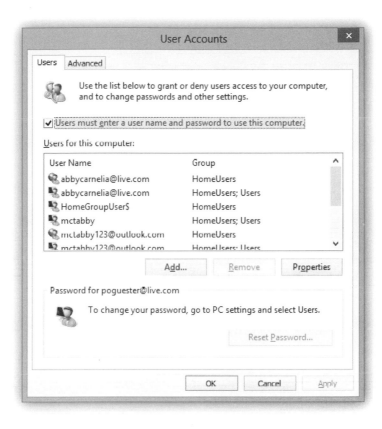

Turn off "Users must enter a user name and password to use this computer." Click OK.

Now tell the PC *who* gets to sign in automatically by entering your account name and password (and the password again); select OK.

Automatically sign in	×

You can set up your computer so that users do not have to type a user name and password to sign in. To do this, specify a user that will be automatically signed in below:

User name:	McTabby
Password:	••••
Confirm Password:	••••|

OK   Cancel

The next time you restart your computer, you'll gasp in amazement as it takes you all the way to the Start screen without bothering to ask for your password.

How to copy files by dragging them

When you drag icons at the Windows desktop, you sometimes *move* them (into a new window), and sometimes *copy* them. Here's the scheme:

- You *move* the files or folders when you drag them to another *folder* on the same disk, and *copy* them if it's a different *disk*.

- If you're pressing the Ctrl key while dragging to another folder on the same disk, you *copy* the icon.

- If you press Shift while dragging to a different disk, you *move* the icon (without leaving a copy behind).

If your reaction is "How am I supposed to remember all that?" you're not alone. Fortunately, you don't have to.

You have to remember only one trick: Use the *right* mouse button as you drag. When you release the button, this shortcut menu appears, offering you a *choice* of what you want to happen: "Move here" or "Copy here." Click the one you want.

When text is too small

Nobody has ever accused Windows of not offering enough settings. Fortunately, some of them are actually useful—and one of them is the ability to *make type larger*. After all, most of us will one day be over 40.

This trick is especially useful because the resolution of computer screens keeps getting higher. Manufacturers keep packing more and more dots into the same space—smaller and smaller dots—and therefore the *type and graphics* are getting smaller.

Here's the quickest fix: Right-click the desktop. From the shortcut menu, choose "Screen resolution" or "Resolution." A dialog box appears; select "Make text and other items larger or smaller."

You wind up here:

Click one of the options here: Medium, Larger, Extra Large; then click Apply. A message now says, "You must log off your computer." Click "Log off now." The next time you turn on the thing, you'll be able to enjoy bigger type and graphics.

If you prefer an in-between magnification, or greater magnification (up to 500%), click "Custom sizing options." Drag the slider until the sample text looks good; click OK. (If you go overboard, you may see blurry type in older programs.)

How to right-click when there's no button

As you know from page 78, *right-clicking* is a hugely important computer skill—especially in Windows.

If you're using a desktop PC, with a two-button mouse attached, figuring out how to right-click shouldn't take long. But if you have a laptop—what then? It has a trackpad, not a mouse.

Laptop makers have solved this problem in different ways. Sometimes you get "mouse buttons" below the trackpad.

But other trackpads may have nothing resembling two side-by-side mouse buttons. In those situations, here are some tips:

- On most laptops, you can click by just *touching* the trackpad; you don't have to fully click down on it.

- You can usually trigger a right-click by tapping or clicking the trackpad with *two* fingers.

- There's often an area on the trackpad that's dedicated to right-clicking—the right half or the lower-right corner, for example, as shown here.

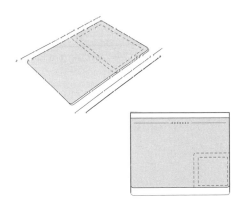

Oh—and what if you're using a Windows 8 touch screen? In that case, you can "right-click" by touching the screen and holding your finger there for a moment.

The power of Backspace

Your Backspace key is useful for more than just deleting stuff you've typed. When you're *not* typing, it has a secondary meaning: Go Back.

That trick works in your Web browser (like Internet Explorer, Firefox, or Chrome) and at the desktop (known as Explorer windows). If you've drilled down several folders deep, for example, you can tap Backspace to "walk backward" out again.

Put another way, the Backspace key takes you "back a space"—get it?

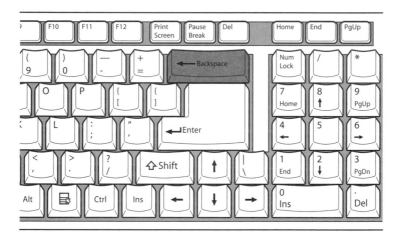

Take a picture of the Windows screen

A *screenshot* is a graphics file that captures the current picture of the screen. Screenshots are great when you want to report a bug, error message, or visual glitch to somebody, or when you want to preserve something really cool or important on the Web, for example.

- **To capture the screen to your Clipboard:** Press the PrtScn key. You've just copied the picture of the screen to your invisible Clipboard; you can now paste it somewhere (into an outgoing e-mail, for example).

- **To capture only a window:** If you hold down the Alt key as you tap PrtScn, you capture the image only of the frontmost *window*, trimming away the rest of the screen.

- **To save the screen image as a file:** If you add the Windows logo key (⊞ or ⊞), tapping the PrtScn key deposits a new graphics file into a folder called Screenshots, in your Pictures folder. (That's instead of putting it on your Clipboard.)

Finally, it's worth noting that recent Windows versions come with a program called Snipping Tool. It's more flexible than the PrtScn key, because it lets you drag freeform shapes for your capture; it lets you edit the image before you save it; and it lets you choose the graphics format for saving.

To try it out, select Snipping Tool on your Start screen (or Start menu, if you have one).

Close a window from the keyboard

The most obvious way to close a Windows window is to click the little x button in the upper-right corner:

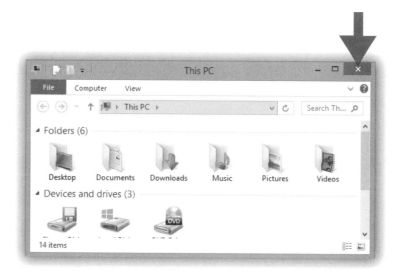

But there's a keyboard shortcut, too: Alt+F4.

A keyboard shortcut is more useful than a button—especially when some cascade of windows has just rained in on your screen. You can just hammer away on the same keystroke over and over, rather than trying to chase a sequence of Close buttons in different places.

The Notepad's handy date stamp

You might think that Notepad, a Windows accessory since 1985, has long ago been surpassed by other, better programs. And yes, that's probably true—but it's still around, and people still use it for jotting down notes, phone numbers, addresses, driving directions, and so on. (If you have Windows 7 or earlier, it's in the Accessories folder of your Start menu; if you have Windows 8 or later, type *notep* at the Start screen to find it.)

Notepad doesn't have a lot of features, but it has one that can be very useful: automatic date stamps. It's often convenient to let Notepad record the date and time of each new entry or brainstorm:

- **Type out the current date and time on command** by pressing the F5 key. Notepad inserts the time stamp wherever your insertion point happens to be.

- **Auto-insert the current date and time every time you open it.** This trick offers a handy way to create a record of when you last worked on a Notepad file—great for diaries, spending records, and so on.

  To set this up, create a new Notepad file (open the File menu and choose New). Type *.LOG* as the first line of the document. (Capitalize *LOG* and put nothing, not even a space, before the period.)

  Save and name the document (from the File menu, choose Save) wherever and whatever you like. Notepad adds the file name extension *.txt* automatically.

From now on, every time you open that Notepad file, you'll see the date and time inserted at the top—and your cursor helpfully deposited on the next line. Now you're ready to type the day's entry.

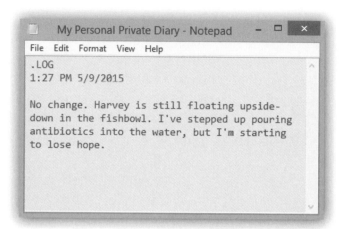

(You'll probably want to press Enter, inserting a blank line, after each entry, before saving the file; your log will be easier to read.)

The "Open Start menu" key

The Start menu (or, in Windows 8, the Start screen) is your home base. It's the first stop for any task. And if you have a mouse or trackpad, it's a pretty small target—a little spot in the lower-left corner.

Fortunately, the ▲ or ⊞ key on your keyboard offers a speedy shortcut: You can tap it to open the Start menu or return to the Start screen.

Restoring the Start menu

In Windows 8, Microsoft took away the most famous element in all of Windows Land: the Start menu in the lower-left corner. Across the country, that change prompted much wailing, gnashing of teeth, and rending of garments.

Without the Start menu, it's much harder to find the programs you want to open—and to access important features like the Control Panel and the Shut Down command.

In Windows 8.1, Microsoft restored a ∎ button to the lower-left corner of the screen, but it's still not the Start menu. That button just opens the Start *screen*: an endless horizontally scrolling world of big square tiles that represent your programs.

Fortunately, restoring the *real* Start menu is a quick download away. Just download and install a free app like Classic Shell (*www.classicshell.net*).

The Secret Start menu of Windows 8.1

Even without installing an app that gives you back the Start menu, as described in the previous tip, Microsoft didn't throw common sense to the wind *completely*. Starting in Windows 8.1, you can open a *hidden* Start menu that offers direct access to a list of important places and programs.

To open the hidden Start menu, *right-click* the ■ button.

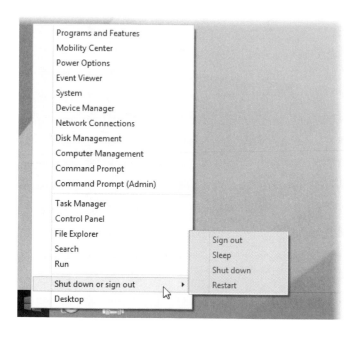

The ▇ button is always visible when you're at the standard Windows desktop. If you're at the Start screen—the tiles—it appears only when you move the mouse to the lower-left corner. Or, on a touch screen, when you swipe in from the left edge of the screen.

Or, no matter where you are, the secret Start menu opens when you press ▇+X.

In any case, this isn't the traditional Start menu. It doesn't list any of your programs, for example. But it *does* list things like Control Panel, Search, and Shut Down—all frequent commands that would otherwise require more steps to find.

Type-searching in Windows 8

In TileWorld, the Start menu no longer exists; it's become a Start *screen*. Each tile represents an app, a file, or a Web page.

Each tile isn't just a button that *opens* a program. It's also a tiny billboard that displays up to date information from that program. The Mail tile shows the subject line of your newest message. The Calendar tile displays the name of your next appointment. The People tile shows incoming posts from Twitter and Facebook.

Once the Search screen gets to be many screens wide, scrolling horizontally to find an app becomes exhausting. But if your computer has a physical keyboard, you can exploit the world's greatest type-selecting shortcut: Just *start typing* the name of the tile you want. (You don't have to click Search or do anything else first.) The screen instantly changes to show you nothing but the icons of matching items.

Enlarging the cursor

There are two reasons you might want to make your Windows cursor bigger. First, each year, computer-screen models come with higher resolution (pixels packed into each inch), so everything on the screen gets smaller.

Second, you might be getting older.

When you make the cursor bigger, nothing else is affected—only your ability to spot it on a big screen.

Start by right-clicking the desktop. From the shortcut menu, choose Personalize. In the resulting dialog box, click "Change mouse pointers" at the left side.

Now you're treated to the Mouse Properties dialog box, shown here:

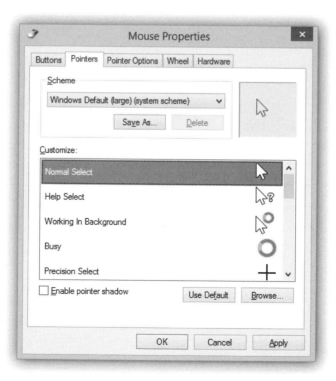

See the Scheme pop-up menu near the top? Open it. Inside, you'll find options called "Windows Default (large) (system scheme)" and "Windows Default (extra large) (system scheme)." Those two options make the Windows cursors bigger and much bigger, respectively. Choose the one you want, then click OK.

Command your PC by voice

Maybe you've heard of Dragon NaturallySpeaking, the program that translates your speech into typing. But Windows has a very similar feature built right in; it lets you both speak to type *and* control the computer, as if you were clicking buttons and opening menus, by voice. You might be amazed at how easy it is—and how accurate it is.

To try it out, open Windows Speech Recognition:

- **Windows 7:** From the Start menu, choose Control Panel. Click Ease of Access, and then click Speech Recognition.

- **Windows 8:** At the Start screen, type *speech*. Select Windows Speech Recognition in the list of results.

Click your way through the screens (click Take Speech Tutorial if you're offered one). Along the way, you'll specify what kind of microphone you have (a headset works best); you'll be able to print a cheat sheet of commands; and you'll be treated to an excellent 30-minute tutorial in which you practice operating your PC by voice.

When it's all over, the Speech palette appears, and you can start talking. The most important spoken commands are these:

- **"Start listening"/"Stop listening."** Say "Start listening" to turn on your mike—you see the microphone button on the Speech palette darken. Say "Stop listening" when you want to speak to a person or your phone. (Ctrl+⊞ is the keyboard shortcut.)

- **"What can I say?"** You get the Speech Recognition page of the Windows Help system, complete with a collapsible list of the things you can say.

- **"Start Word."** Opens the program you've named. "Start Calculator." "Start Internet Explorer." Whatever.

- **"Switch to Word."** Switches to the program you've named.

- **"File; Open."** You can control menus by saying whatever you would have clicked. If you say "Edit; Select All," it's as though you had opened the Edit menu and picked Select All.

- **"Print."** You can also "click" any button on the screen, or any link on a Web page, by saying it: "OK," "Cancel," and so on.

- **"Double-click Recycle Bin."** You can "double-click" or "right-click" anything you see—by voice.

Type by talking to your PC

A s you now know, you can command your PC by voice—but you can also speak freely, *dictating* text instead of typing it. You know: "Dear John (comma; new line), How long have I known you (question mark)? I'm afraid that the answer is (comma), too long (period)."

(Yes, you have to speak the punctuation.)

You might be amazed at how fast and accurate the program is.

Now and then, Speech Recognition makes an error. In that case, you should correct it by voice, so it learns from its mistake. Suppose, for example, that you said "oxymoron," but Windows typed, "ax a moron."

In that case, say, "Correct *ax a moron*." When the list of alternative transcriptions appears, say the number of the corrected interpretation ("three"). If you don't see it, speak the correct text again. (In a pinch, you can also say, "Spell it," and then spell it out loud.)

Once you've found (or said) the correct phrase, say "OK" to close the panel, replace the corrected text, and teach Windows not to make that mistake again.

Keystrokes for the Two Worlds of Windows 8

If you have Windows 8, you live in two worlds. First, there's the regular Windows desktop that's been around forever (facing page, top). It evolved along with the mouse and keyboard, and with them, does a great job.

Then there's TileWorld, born in 2012. It's the Start-screen world made of colorful tiles (facing page, bottom). TileWorld was created for the new era of touch screens.

Unfortunately, now you're stuck with two different software worlds. Fortunately, you can flip back and forth fairly easily—from the keyboard, if you like.

- **Jump into TileWorld:** Press the ⊞ key on your keyboard. (On a Windows 8 tablet, there's a ⊞ button instead of a key.) If you press the ⊞ key again, you return to whatever

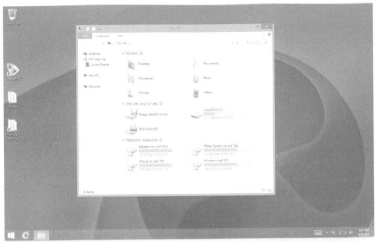

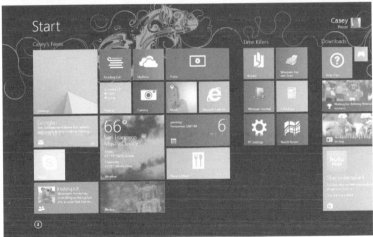

app you were just using. In other words, presssing ⊞ repeatedly flips back and forth between the app and the Start screen.

- **Jump to the desktop:** Click the Desktop tile. Or press ⊞+D (for *desktop*, get it?).

But here's another trick for getting to the desktop with even less physical effort.

It turns out that at the Start screen, pressing the Enter key always opens the *top left tile*. And *you* get to decide what to put there.

If you're smart, you'll drag the *Desktop* tile there, like this:

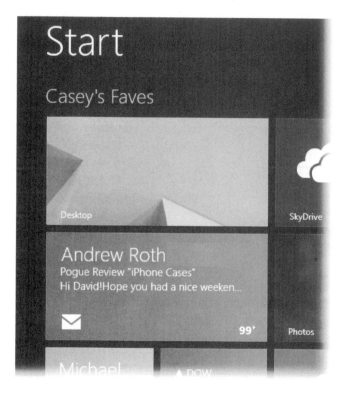

Why? Because now, from now on, you have single keys to jump between the two worlds of Windows. Tap ▦ for the Start screen, and press Enter for the desktop.

That's right, a big fat key for each world. Because, you know, the mouse is for sissies.

How to read file names that are cut off

If a Windows window contains files whose names are very long, they'll get chopped off. Sometimes, the end of a file name is important; it can make the difference between "Recipe for Delicious Lemony-Filled Home-Baked Donuts" and "Recipe for Delicious Rat-Killer Poison Cakes."

When that happens, you have two options:

- **Point to the name without clicking.** After a moment, a pop-up label appears, showing you the entire name.

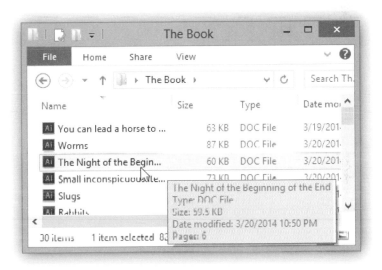

- **Double-click the little line that divides one column title from the next.** You can see the effect illustrated on the next page.

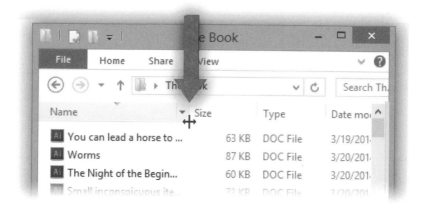

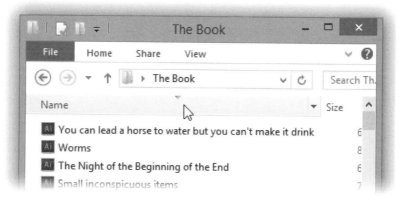

When you do that, the *entire column* widens exactly enough to show you the full column names without chopping them off. (You can always drag that column smaller again, using the same divider line as a handle.)

How to rename a whole bunch of files in one fell swoop

Renaming a file or folder in Windows is pretty easy. To open the little rectangle where you can rename it, you can either:

• Click the icon and then press the F2 key.

• Click the icon twice, slowly.

But here's something not many people realize: You can rename a whole *bunch* of icons all at once.

Just select them all, using the tip on page 90. Then press F2.

Now, rename *one* of the files, as shown below at left. When you press Enter or click somewhere else, *all* of the icons now have the same name.

If there is more than one icon of the same type (folders or documents, say), Windows helpfully adds a (1), (2), and so on to their names.

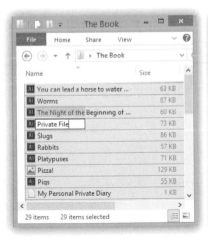 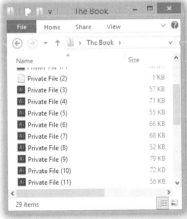

One-key screen locking before fetching coffee

I f you use your computer in a place where other people also live or work—it could happen—you'll like this one.

With a single keystroke, you can lock your screen. You can get up to go get coffee, stretch your legs, or un-jam the printer, without worrying about coworkers or evildoers seeing whatever confidential Web site or document you've left open.

And that keystroke is ⊞+L (for "Lock," get it?).

3:17
70°
Washington D.C.
Mostly Sunny
84°/64°
Friday, May 9

Once your PC is locked, nobody can get into your stuff again without your password.

The top 10 things the Windows-logo key (⊞) does on every PC

Most of the keys on the standard PC keyboard are descended from their great-grandparents on the typewriter. You probably know what's going to happen if you press the M key or the 1 key.

A few, though, have never graced a single typewriter—and the Windows-logo key (⊞ or ⊞) is one of them. But it's worth meeting, because its powers are many.

To do this	Press this key
Open the Start menu or Start screen	⊞
Display the desktop	⊞+D
Lock the tablet screen in current orientation	⊞+O
Minimize all windows	⊞+M
Restore minimized windows to the desktop	⊞+Shift+M
Open an Explorer (desktop) window	⊞+E
Lock your computer or switch users	⊞+L
Open the Run dialog box	⊞+R
Cycle through programs on the taskbar	⊞+T
Switches among monitors or projectors	⊞+P

12 important things
the ⊞ key does in Windows 8

In creating Windows 8 (and 8.1, and so on), Microsoft gave the humble ⊞ key even more importance. Now it's extremely helpful in navigating the new, touch-screen-friendly, Start-screen world of jumbo tiles.

Open the Start screen	⊞
Open secret Utilities "Start" menu	⊞+X
Open Charms bar	⊞+C
Open App (options) bar	⊞+Z
Search everything	⊞+Q or ⊞+S
Search for files	⊞+F
Search for settings	⊞+W
Open the Share panel	⊞+H
Open Devices panel	⊞+K
Open Settings panel	⊞+I
Cycle through open tile-based apps	⊞+Tab
Cycle backward through open tile-based apps	⊞+Shift+Tab

Taming the Recycle Bin

The Recycle Bin, of course, is where you toss files and folders that you don't want anymore. It's not so much a trash can as a *waiting room* for the trash can, because things generally don't disappear from it until you *empty* it.

You can drag files or folders onto the Recycle Bin icon, but it's usually faster to highlight them and then press the Delete key.

Ordinarily, there's another step: Windows asks you if you're *sure* you know what you're doing. Since nothing's actually deleted yet anyway, you may as well turn *off* this warning window.

To do that, right-click the Recycle Bin icon; from the shortcut menu, choose Properties. In the Properties dialog box turn off "Display delete confirmation dialog." Now you'll never get that message when you put something into the Recycle Bin.

Once you've put something in the Recycle Bin, one of three things will happen:

- **You'll change your mind.** You'll rescue the file from doom. You'll double-click the Recycle Bin icon to open its window. You'll right-click the reprieved icon and, from the shortcut menu, choose Restore. The file flies back to the folder from whence it came. (Or just drag the icon out of the Recycle Bin window into any other window or folder.)

- **You'll proceed with termination.** Right-click the Recycle Bin icon (or any empty spot in the Recycle Bin window). From the shortcut menu, choose Empty Recycle Bin. In the "Are you sure?" message, click Yes.

- **The Recycle Bin will empty itself.** When the Recycle Bin gets so full that it's occupying more than 10 percent of your disk space, it will start auto-deleting older Recycle Bin files as new ones arrive.

If that notion alarms you, all is not lost. Right-click the Recycle Bin's icon; from the shortcut menu, choose Properties. In the resulting dialog box, you can change the "Maximum size" number. Keeping the number low means that you're less likely to run out of disk space. However, a high number increases the number of files you'll have a chance to rescue if you change your mind.

A keyboard shortcut
for anything

Keyboard shortcuts are efficient and satisfying to use, but you don't want to go overboard.

Just kidding. There's no such thing as going overboard!

That's why it's very cool that Windows lets you set up keyboard shortcuts for *any file, folder, and program on your computer.* Press the keystroke to open that thing without lifting your hands from the keyboard.

There is some fine print, however, as you'll discover when you follow these steps:

1. **Make a shortcut of the icon.** A *shortcut* is like a duplicate icon for a file, folder, disk, or program—but it occupies almost no disk space. When you double-click the shortcut icon, the original icon opens. Shortcuts, in effect, let you keep a certain item's icon in more than one place. And the make-your-own-keystroke feature works only on shortcuts, not on original icons.

   To make a shortcut of an icon, right-click it; from the shortcut menu, choose "Create shortcut." The new icon appears right next to the original. You have to put it on your desktop for this trick to work, though; it can't be in a folder.

2. **Open the shortcut's Properties box.** That is, right-click its icon; from the shortcut menu, choose Properties.

3. **Click in the Shortcut Key box and press the keyboard combo you want.** Your combination *must* include Ctrl+Alt, Ctrl+Shift, or Alt+Shift, and another key. And it can't include the Space bar or the Enter, Backspace, Delete, Esc, Print Screen, or Tab keys.

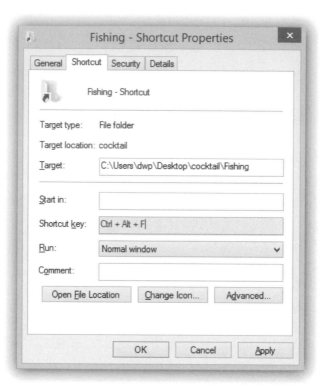

4. Click OK. Now try your magic keystroke! And marvel that, just this once, *you* are the master of the PC instead of the other way around.

How to type symbols

The typical PC's keyboard sure has a lot of keys. But you won't find useful symbols like ©, ™, ÷, and ¿ on any of them.

In Microsoft programs like Word, Excel, and PowerPoint, a handy palette of them is available when you open the Insert menu and choose Symbol. Click the one you want, then click Insert.

In other situations, you can use the Character Map program. Find and open it as you would any other program (type it into your Start menu search box, or type it at the Windows 8 Start screen).

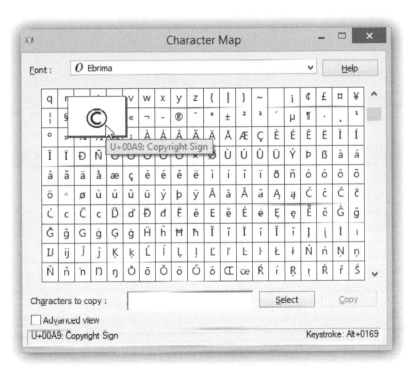

When it opens, use the Font pop-up menu to specify the font you want to use; every font contains a different set of symbols. Double-click a character to transfer it to the "Characters to copy" box. Click Copy, and then Close. When you return to your document, use the Paste command to insert the symbol you've chosen.

Blow past the Lock screen

When you turn on a Windows 8 PC, you're greeted by a decorative and mostly useless Lock screen. Like the one on a smartphone, it shows you the time, date, Wi-Fi signal strength, weather, and (on laptops and tablets) battery charge.

On a desktop computer, though, the Lock screen is just a layer of red tape.

Fortunately, blowing past it is easy; almost anything you do gets rid of it. Press any key, click anywhere, turn the mouse wheel, or (if you have a touch screen) swipe upward on the glass.

Chapter 8:
Word Processing, Number Crunching, Slideshowing

Microsoft Office is the set of programs that towers over the world's corporate workers like Godzilla. It's available in similar versions for both Mac and Windows.

Office comes in different versions, but all of them include the Big Three: Word (for word processing), Excel (for spreadsheets and number crunching), and PowerPoint (for business slideshows). All three are *crawling* with secrets.

The five very, very basic keystrokes worth learning

When you're in Word, Excel, or PowerPoint, a few important keyboard shortcuts can save you all kinds of time and give you a glowing sense of mastery. Here they are:

Ctrl+S   Save (any changes you've made; press this frequently).
Ctrl+F   Find (some words or numbers in the document).
Ctrl+O   Open (another document).
Ctrl+Z   Undo (the last change you made).
Ctrl+Y   Redo (the change you just *undid*).

On the Mac, press the ⌘ key in place of Ctrl in the list above.

Turn off automatic bullets, lists, and links

It's nice that Microsoft Word is *the* Word Processor. You never have to worry that your colleagues won't be able to open your document; they all own the same program.

What's less nice is how aggressively Microsoft Word tries to format what you type. Suppose, for example, that you're writing a stern note to your kid. You type this:

Chris: As we've discussed, I'm happy to offer you an allowance. But in order to earn that $1.25 a week, I expect you to perform *certain chores around the house. As we've discussed, those tasks in-clude:*

1. cleaning the gutters

Now you press Return to type the second item—and Word does something alarming. It indents your first item and adds a "2". on the next line—without so much as asking you. Now you've got this:

1. cleaning the gutters
2.

Microsoft thinks it's doing you a favor by auto-numbering your list. And that's just the beginning. It changes *phrases like this* into boldface. It turns two hyphens (--) into long dashes (—). When you type an Internet address, it turns it blue and underlined and clickable (http://www.wordisannoying.com).

Millions of people suffer in silence. But you have two weapons of resistance.

First, you can press Ctrl+Z (on the Mac, ⌘-Z) to undo the formatting each time Word does it.

Second, you can turn off this behavior *for good.*

- **In the Mac version of Word:** Open the Tools menu and choose AutoCorrect. Click the "AutoFormat as You Type" tab, as shown here.

- **In the Windows version:** At the top of the screen, click File. Then click Options. Then Proofing. Then AutoCorrect Options.

In both cases, you can now start turning off the checkboxes of the things that annoy you, and then click OK.

Auto-column widths in Excel

Excel is a spreadsheet: a grid of rows and columns intended for analyzing and processing numbers.

Clearly, you want the columns to be as wide as necessary to show you everything in them—but no wider, so you can fit as much information onto the screen as possible.

You can adjust their widths manually—if you have all day— by dragging the tiny divider line at the *right* side of any column's identifying letter, as shown here at top:

	A	B	C	D	E
1	Date	PVs	TS	UU	Bcookie
2	7-Jan	364,172	472,766	154,768	186,931
3	8-Jan	2,310,786	3,905,782	1,261,016	1,520,285
4	9-Jan	1,790,357	4,087,136	966,342	1,163,088
5	10-Jan	1,457,546	3,476,975	821,964	988,470
6	11-Jan	994,046	1,756,499	603,314	723,926
7	12-Jan	502,714	544,614	279,720	336,391

	A	B	C	D	E	F	G	H	I
1	Date	PVs	TS	UU	Bcookie	PV/UU	TS/UU	CTR	Share
2	7-Jan	364,172	472,766	154,768	186,931	2.35	3.05	0.41%	32
3	8-Jan	2,310,786	3,905,782	1,261,016	1,520,285	1.83	3.10	0.24%	8,6
4	9-Jan	1,790,357	4,087,136	966,342	1,163,088	1.85	4.23	0.25%	8,70
5	10-Jan	1,457,546	3,476,975	821,964	988,470	1.77	4.23	0.23%	6,44
6	11-Jan	994,046	1,756,499	603,314	723,926	1.65	2.91	0.22%	4,7
7	12-Jan	502,714	544,614	279,720	336,391	1.80	1.95	0.30%	1,14

Excel is happy to do that job automatically. Position your pointer on that divider line—and double-click. Presto: Excel auto-adjusts the width of the column to exactly accommodate the widest cell.

You can auto-widen lots of columns at once, too. Just drag through the column titles (A, B, C, and so on) to highlight them, and then double-click any *one* of the divider lines (as shown at bottom).

Jump among Excel sheets

A n Excel spreadsheet may actually be spread*sheets*. Take a look at the bottom edge. If you see several tabs, then you've got several spreadsheets stapled together into a single document.

That's the first tip—just to be aware. More than one baffled coworker has stared at the first spreadsheet page, trying in vain to find numbers that are supposed to be there, unaware that they're *behind* that first sheet on one of the other tabs.

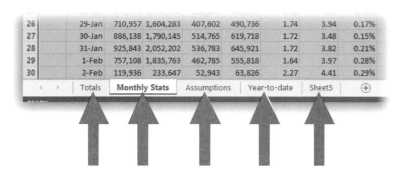

The second tip is to use the keyboard shortcut to switch sheets. While holding down the Ctrl key (⌘ on the Mac), press the PageUp or PageDown key.

Paste without the formatting

Here's a tip for Microsoft programs like Word, PowerPoint, and Outlook: When you paste in text from another document—say, a Web site—you might not want all the boldface, fonts, and other formatting from the original source. It often clashes with the text that's already there, as shown here at top.

Dear mom:

I think the NY Times has finally answered my question about whether or not spanking = child abuse. The article says that "Six years ago the psychologist Elizabeth Thompson Gershoff, then at Columbia University, published a review of 62 years of research, analyzing 82 separate studies. And while there was a lot of evidence that spanking makes children do what they are told in the very short term, it seems only to teach children not to get caught. What it doesn't do is teach them to do better."

Dear mom:

I think the NY Times has finally answered my question about whether or not spanking = child abuse. The article says that, "Six years ago the psychologist Elizabeth Thompson Gershoff, then at Columbia University, published a review of 62 years of research, analyzing 82 separate studies. And while there was a lot of evidence that spanking makes children do what they are told in the very short term, it seems only to teach children not to get caught. What it dooon't do is teach them to do better."

To avoid that problem, open the Edit menu and click Paste Special. Click Unformatted Text. You get just the text, without the fanciness (shown at bottom).

Many other programs have similar commands. In Apple Mail, it's called Paste and Match Style. In Adobe InDesign, it's Paste Without Formatting. You get the idea.

The quick way to edit a spreadsheet cell

When you want to change the number or formula in an Excel cell, you might be inclined to double-click it. That, after all, is how you usually "open it" for editing (shown here at top).

But experienced spreadsheet jockeys also know about the F2 trick.

If you click the cell and then tap the F2 key (top of your keyboard), the cell opens *and* the blinking insertion point appears at the end of whatever's in it (shown here at bottom). You're all set to type something more, or to press Backspace and edit what's there.

C16	fx	Mac Pro Review	
	C		M
12			To
13	**Video Name**		**Video Star**
14	Real Estate of the Rich and Geeky		495,15
15	Google Glass Will Never Be Okay		254,15
16	Mac Pro Review	✛	176,72
17	Demo of Skeletonics, a $50,000 Exoskeleton at SXSW		154,45
18	Yancey Strickler says he will never sell Kickstarter		95,24

C16	fx	Mac Pro Review	
	C		M
12			To
13	**Video Name**		**Video Star**
14	Real Estate of the Rich and Geeky		495,15
15	Google Glass Will Never Be Okay		254,15
16	Mac Pro Review|		176,72
17	Demo of Skeletonics, a $50,000 Exoskeleton at SXSW		154,45
18	Yancey Strickler says he will never sell Kickstarter		95,24

On the Mac, press Control+U for this function.

The panic of Insert mode

You're typing along in Word, and you suddenly realize that something is *seriously* wrong. When you type, you don't push the existing words off to the right, as usual. Instead, what you type *replaces* what's already there, eating it up with each keystroke. It's the freakiest darned thing, and very upsetting.

3.1. I hereby appoint P▮llip McGlass as Executor. Or, if this Executor is unable or unwilling to serve, th▮ ▮I appoint Mandy Lifeboats as alternate Executor.

3.2. My Executors shall have full and absolute power in his/her discretion to sell all or any assets of my estate, whether by public auction or private sale and shall be entitled to let any property in my estate on such terms and conditions as

3.1. I hereby appoint Phillip McGlass as Executor. Or, if this Executor is unable or unwilling to serve, then I appoint Mandy Lifeboats as alternate Executor

3.2. My Executors shall 3.7. I leave my collection of ceramic dolls retion to sell all or any assets of my estate, whether by public auction or private sale and shall

You've somehow entered Overstrike mode, which is not, ahem, one of Microsoft's finest inventions. It does exactly what you've discovered: turns your computer into a text-eating monstrosity, in which each letter you type gobbles up another letter you had *already* typed.

What has happened, probably, is that you accidentally hit the Insert key. (You were probably reaching for the Backspace key right next to it.) The Insert key turns on Overstrike mode.

Fortunately, the Insert key also turns it *off*. So press the Insert key again to return to the land of the sane.

(In recent versions of Word, it's harder to turn on Overstrike mode, thank goodness.)

Fine-tuning PowerPoint positioning

When you're designing a PowerPoint slide, you probably know that you can drag a picture or a text box around with your mouse. PowerPoint makes the object snap into positions it thinks you want, which you might find helpful and might find infuriating.

You might also know that for finer positioning, you can select an object and then tap your arrow keys to move it more precisely. This trick, too, has its downsides; each arrow-key press moves the object a fraction of an inch—not one pixel, as you might imagine.

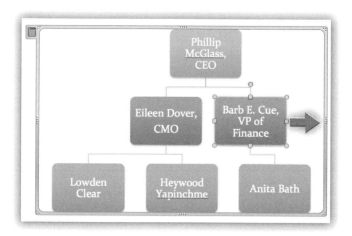

But you probably *didn't* know that if you're pressing the Ctrl key while using the arrow keys, PowerPoint stops trying to outthink you. It moves the object in very, very fine increments, so you can drop it *exactly* where you want it.

Copying down from a cell

As you work on a spreadsheet cell, you may sometimes want to duplicate what's in the cell *above* it—sometimes because you want the two to be identical, and sometimes because you want to edit the lower cell slightly.

In any case, you'll love this: If you click into a cell and press Ctrl+quote mark ("), Excel slaps in whatever's in the cell above it. (It's easy to remember, because the quote mark is like a ditto mark.)

On the Mac, it's the same thing: Control+quote mark.

Handy!

	EAST	WEST
Koyannisquatsi A	176,727	1,957,344
Koyannisquatsi B	154,455	354,006
	95,245	125,443
	58,592	79,111
	57,318	131,127
	50,548	76,677

	EAST	WEST
Koyannisquatsi A	176,727	1,957,344
Koyannisquatsi B	154,455	354,806
Koyannisquatsi B	95,245	125,443
	58,592	79,111
	57,318	131,127

Secrets of the Format Paintbrush

Nobody uses the Format Painter, because nobody understands it. But one demonstration, and you'll be a believer. The Format Painter is a button on the toolbar of Word, Excel, and PowerPoint (shown by the arrow on the facing page). You can use it to transfer the formatting from *one* bit of text—font, size, color, underlining, indentation, and so on—to another. You may have put quite a bit of effort into formatting some paragraph, so the Format Painter can save you an incredible amount of Format Pain.

Here's how it goes:

1. **Select the text that already looks the way you want it.** You can highlight just a word, a whole line, a spreadsheet cell—anything, as shown on the facing page at top.

   If you want *whole-paragraph* formatting transferred—aspects like indentation, bullets, numbering, line spacing, and centering—make sure that your selection also includes the invisible paragraph marker after the last period, as shown on the facing page.

2. **Click the tiny paintbrush on your toolbar—the Format Painter button.** Nothing happens yet. But your cursor has sprouted a + symbol. It's loaded.

3. **Select the text that isn't yet formatted correctly.** And just like that, the new text takes on the formatting of the first (facing page, bottom).

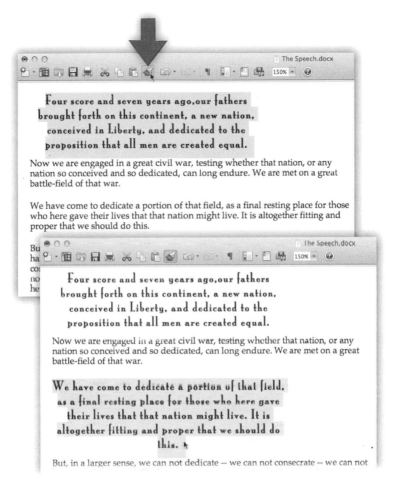

You can save even more time when there are *several* new blobs to be formatted. In that case, in step 2, *double-click* the Format Painter button. Now it's never *off*. You can sweep through your document, joyfully dragging through text or cells, reformatting as you go, without ever having to re-click the Format Painter button.

When you're good and done with the reformatting, press the Esc key to turn off the paintbrush.

Bonus tip: If you forgot to double-click in step 2, all is not lost. Once you've reformatted *one* blob of text, you can highlight another and then press Ctrl+Y (on the Mac, ⌘-Y), which means "repeat whatever I just did." The new text takes on the formatting, without the Format Painter's even knowing about it.

Navigate Word with the keyboard

In the early days of computing, the *only* way to move around the screen was to use keystrokes. Then the mouse was invented, and everybody cheered—much easier, right?

But the thing is, keyboard control has a lot to say for itself. It can be faster and more comfortable than taking your hand off the keyboard and reaching for the mouse. It can be more precise: A single keystroke can jump *directly* to the next word, line, or paragraph.

If you're persuaded, here's the scheme of keyboard navigation in Word. (On the Mac, substitute the ⌘ key for the Ctrl key.)

To move to...	*Press this*
The previous or next letter	← or → key
The previous or next word	Ctrl+← or → key
Previous or next paragraph	Ctrl+↑ or ↓
The beginning or end of line	Home or End
Beginning/end of document	Ctrl+Home or Ctrl+End

Up or down a screenful	PageUp key, PageDown key
Up or down a page	Ctrl+PageUp, Ctrl+PageDown

OK, that's all great for moving the insertion-point cursor around. But here's the megatip:

Add the Shift key to any of those to *select* text as you go. Here's how it works:

Four score and seven years ago,our fathers nation, conceived in Liberty, and dedicated created equal.

Ctrl+→   Four|score and seven years ago,our fathers nation, conceived in Liberty, and dedicated created equal.

Ctrl+→   Four score|and seven years ago,our fathers nation, conceived in Liberty, and dedicated created equal.

Ctrl+→ with Shift key   Four score and seven years ago,our fathers nation, conceived in Liberty, and dedicated created equal.

Four score and seven years ago,our fathers nation, conceived in Liberty, and dedicated created equal.

Four score and seven years ago,our fathers nation, conceived in Liberty, and dedicated created equal.

You can use these tricks in Excel, too. For example, the "end of document" keystroke (Ctrl+End) takes you to the very last cell of your spreadsheet—and adding Shift *highlights* all of the cells to the very last one.

Changing the case,
the quick and easy way

IT'S NOT SO MUCH FUN READING THINGS WRITTEN IN ALL CAPITALS, IS IT? ONLINE, THEY'LL HATE YOU FOR THIS. IT FEELS LIKE YOU'RE SHOUTING.

BUT WHAT IF YOU'VE TYPED A COUPLE OF PARAGRAPHS IN ALL CAPS BY MISTAKE?

You don't have to delete it and retype it all. In Word, Excel, and PowerPoint, a secret keystroke lets you convert the capitalization of any highlighted text.

Ready? It's Shift+F3. (On a Mac, you'll probably have to press the Fn key, too; see page 143.)

Each time you press that keystroke, you cycle among three capitalization styles, shown here in order: All capitals, all lowercase, and sentence case (first word capitalized).

"MY GOD," SAID BOB. "I HAD THE CAPS LOCK KEY DOWN FOR SIX PAGES!"

"my god," said bob. "i had the caps lock key down for six pages!"

"My god," said bob. "i had the caps lock key down for six pages!"

(There is also, by the way, a command in the Format menu called Change Case. It offers even more variations of capitalization. But it's not as quick and slick as the keystroke.)

The quick way to e-mail
an open document

What do you do with a document once you're finished with it?

Maybe you print it. Maybe you save it. Maybe you delete it as soon as possible.

But often, what you want to do is *e-mail it* to someone—a boss, an editor, a colleague.

Microsoft realizes that. So each of its programs—Word, Excel, PowerPoint—offers a "Send as Attachment" command *right in the program.* You don't have to save it, name it, find it, or attach it to an e-mail message.

So: Once you're finished editing a document, open the File menu. From the Share or Send To submenu, choose "E-mail (as attachment)."

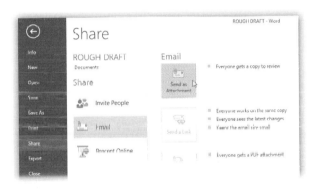

Like a good little minion, your computer automatically opens up your e-mail program (Outlook or Apple Mail, for example), creates an outgoing message, and attaches the document you worked on. All you have to do is address it, add a cheerful message, and click Send.

Split the screen

In a spreadsheet, you can see right away why you might want to split your screen into separately scrolling halves: so that you can remember the column names as you scroll down.

But it can also be useful to split a Word document this way—when you want to refer back to something earlier in the document as you write:

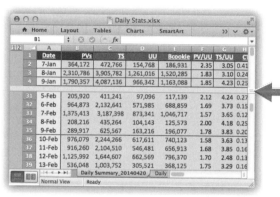

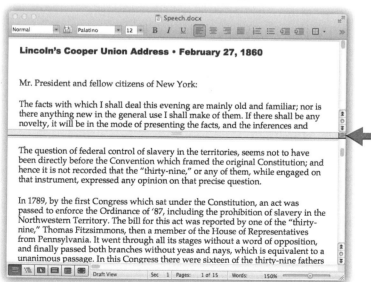

Splitting your window like this is easy. At the top of the *vertical* scroll bar on the right, a tiny, special, draggable handle appears. Grab it and drag it downward to split the window. You can now scroll each half independently.

There is, of course, a keyboard shortcut for this. In Windows, press Alt+Ctrl+S. On the Mac, it's Option-⌘-S. In both cases, the same keystroke unifies the window once again.

What the squiggly underlines mean

While editing your Word, Excel, or PowerPoint documents, you've probably seen the red wavy lines that appear under certain words. That's Microsoft's gentle way of pointing out questionable spellings.

Microsoft isn't saying you've written something *wrong* necessarily—only that you've typed a word it doesn't recognize.

The beauty of this system is that you usually don't even have to know the correct spelling or grammar rule. Just *right-click* the underlined word (page 78). Like magic, the software offers a pop-up menu of correctly spelled suggested replacements. Choose the correct option to fix the spelling.

We have come to dedicate a porcion of that field, as a final
resting place for those who here gave their lifes that that
nation might live. It is altogether fiting and proper that we
should do this.

    fitting
    feting

But, in a larger sense, we can not dedic fating
consecrate — we can not hallow — this fisting
men, living and dead, who struggled h fêting
consecrated it, far above our poor pow
The world will little note, nor long rem Ignore
here, but it can never forget what they Ignore All
    Add

It is for us the living, rather, to be dedi AutoCorrect ▶
unfinished work which they who foug Spelling...
so nobly advanced. It is rather for us to be here dedicated to
the great task remaining before us -- that from these
honored dead we take increased devotion to that cause for

Other times, you'll know that what you wrote was correct
the first time. (Maybe you're writing a novel about somebody
with bad spelling, and you're misspelling words on purpose.)

In that case, right-click the word; from the shortcut menu,
choose Ignore (this occurrence of the word), Ignore All (occur-
rences of the word in this document), or Learn (this word, so it's
considered correct from now on).

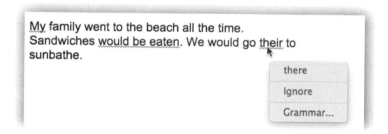

My family went to the beach all the time.
Sandwiches would be eaten. We would go their to
sunbathe.

    there
    Ignore
    Grammar...

And when you see *green* squiggles, that's Word's grammar
checker finding something it doesn't like. Maybe it's a word you
typed twice accidentally, for example, or you used passive voice.
The fix is the same: Right-click to see Word's suggestions.

The logic of the Z, X, C, V keystrokes

If you're using a computer instead of a typewriter, it may be because it's so much easier to *edit* your work. You can cut and paste sections and move them around.

If you do a lot of that, then it's worth learning the shortcuts:

- **Cut** is Ctrl+X (or ⌘-X on the Mac)

- **Copy** is Ctrl+C (or ⌘-C on the Mac)

- **Paste** is Ctrl+V (or ⌘-V on the Mac)

- **Undo** is Ctrl+Z (or ⌘-Z on the Mac)

These are pretty easy to remember. X looks like a pair of scissors *Cutting;* C stands for Copy; V points downward, where you're laying the copied material (or maybe looks like an upside-down "insert" proofreader's mark).

And Z? Well, it's the last letter of the alphabet—and the keystroke undoes the *last* thing you did.

It's kind of cool that on your keyboard, all four of those letters sit in a row. And they're conveniently installed *just* above the Ctrl or ⌘ key, too.

Excel's automatic list maker

You can probably think of many times when you might want to create a list of sequential cell labels in a spreadsheet. Maybe it's Jan., Feb., Mar.... or Mon., Tues., Weds.... or 2015, 2016, 2017.

Excel can do the typing for you.

Start it out by manually filling in the first *two* cells of the series—enough to establish the pattern. Highlight them.

Then grab the tiny handle in the corner of the last cell—and drag it. In any direction. As you drag across other cells, Excel fills them in automatically, continuing your sequence.

It's incredibly smart. If you filled in 1 and 3, it knows you want the following cells to go 5, 7, 9, 11. If you filled in *Lesson 1* and *Lesson 2,* it knows to fill in *Lesson 3, Lesson 4, Lesson 5,* and so on. If you filled in 3:00 p.m. and 3:15 p.m., it fills in 3:30 p.m., 3:45 p.m., and so on.

The time you save could be your own.

Press the B key to black out a slide

If you're giving a presentation in PowerPoint or Keynote, press the B key to black out the slide. That's handy when you want your audience to pay attention to *you* instead of your slide. Press B again to bring the picture back.

The W key works, too—to *white* out the slide.

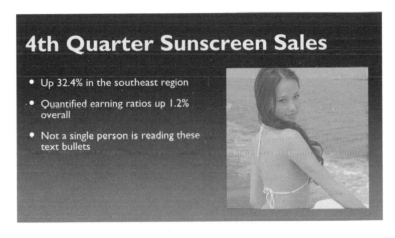

Start the slideshow
—right now

When it's time to start your PowerPoint slideshow, don't bother hunting for a command or a little icon; just press the F5 key. The slideshow starts at the beginning.

And if you don't *want* the beginning, don't embarrass yourself by starting at the beginning and frantically clicking your way back to the slide you really want. Instead, press *Shift*+F5. PowerPoint hops into presentation mode—beginning with the slide you had up.

Jump to a certain slide

You're at the end of your PowerPoint slideshow, taking questions, and somebody asks you about a point you made back on slide 6. Or you're giving your presentation, and you realize that you need to skip a slide.

Either way, it's useful to be able to leap directly to a certain slide without making your audience watch you flip through all of them in sequence.

Type a number . . .   . . . to jump to that slide

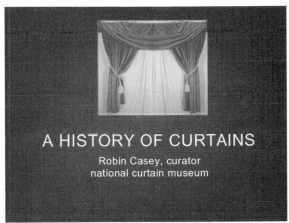

All you have to do is *type the slide number* and then hit Enter. You don't have to exit slideshow mode, and you won't see your typing on the screen. Just type the number 1 (and then hit Enter) to jump back to the start, or 17 (and Enter) to go to that slide, and so on.

The free versions of Microsoft Office

M ost of the business world pays for Microsoft Office. Year after year. Billions of dollars.

But it's perfectly possible to create, open, and edit Word, Excel, and PowerPoint documents without every buying Microsoft Office. All you need is one of the *free* alternative suites, like Apache OpenOffice and Libre Office. (You can get them right now from *www.openoffice.org* and *www.libreoffice.org*.)

Those are full-blown, full-featured suites—not as polished-looking as Microsoft Office but plenty powerful for what most people do most of the time.

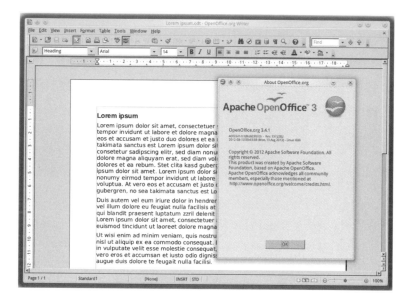

Don't miss Google Drive, either (formerly Google Docs). It's a free *online* set of programs that let you work with Word, Excel, and PowerPoint files. You can try it out at *drive.google.com*.

Chapter 9:
What Not to Do

Anything as complex and powerful as a computer inevitably comes with a few booby traps. Armed with a computer and an Internet connection, you can get into all *kinds* of trouble.

It's no surprise that companies like DriveSavers do a thriving business recovering files from hard drives that have crashed. And not *much* of a surprise that DriveSavers employs a former suicide-hotline counselor.

It's not so difficult to avoid letting your computer ruin your whole day. You just need to be aware of where the mines are buried.

How to delete an entire document with one keystroke

You may remember from page 84 that once you've high-lighted some text, typing anything *replaces it.*

That's supposed to be a convenience. It saves you the step of pressing Delete or Backspace before typing the replacement text.

It's also a Black Pit of Hell for people who select text *accidentally* and then keep typing. Suddenly a whole lot of great work and effort is lost—in the blink of an eye.

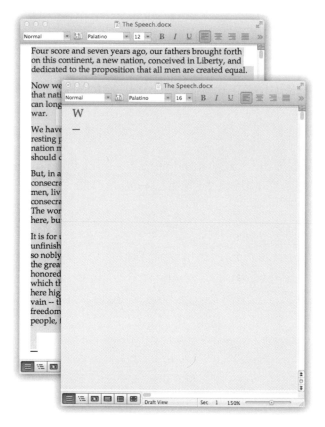

Especially if what you selected was *everything*—the entire document. Select all, type *one single key,* and your entire 70-page term paper is vaporized.

You do, however, have a safety net: the Undo command. As soon as you discover what you've done, press Ctrl+Z (that's ⌘-Z on the Mac). More than once, if necessary, to bring back the text you deleted.

How to have your identity stolen and your life complicated

It's one of the most common scams on the Internet, and thousands of people fall for it every day. It works like this:

You get an e-mail from a big company—maybe Apple, PayPal, Yahoo, or your bank. It reports that there's a problem with your account. You click the link to see what's up. You go to the company's Web site, you log in—and the damage is done. You've just given the bad guys your name and password. Your account now belongs to them.

The e-mail was fake. (Sometimes, bad spelling or grammar or typography gives you a clue.)

The Web site it opened also was fake. It was designed to look exactly like your bank's Web site (or PayPal, or Apple, or whatever). When you "logged in" with your name and password, the bad guys intercepted it.

This scam is known as *phishing* (because they're "fishing" for your information, get it?). Don't fall for it. *Real* banks and companies never send e-mails like that—only crooks.

If you ever wonder if there really *is* a problem with your bank, PayPal, or Apple account, *don't click the link in the e-mail.* Instead, open your Web browser yourself. Go to the company's Web site the usual way (not by clicking a link in an e-mail). Then log in normally.

In most e-mail programs, furthermore, you can see if the link is fishy—by pointing to it without clicking. If you see that the link doesn't match the *underlying* Web address, as shown here, don't click.

From: PayPal Billing Department <Billing@PayPal.com>
Subject: **Credit/Debit card update**
Reply-To: Billing@PayPal.com

PayPal

Dear Paypal valued member,

Due to concerns, for the safety and integrity of the paypal account we have issued this warning message.

It has come to our attention that your account information needs to be updated due to inactive members, frauds and spoof reports. If you could please take 5-10 minutes out of your online experience and renew your records you will not run into any future problems with the online service. However, failure to update your records will result in account suspension This notification expires on 48.

Once you have updated your account records your paypal account service will not be interrupted and will continue as normal.

Please follow the link below and login to your account and renew your account information

https://www.paypal.com/cgi-bin/webscr?cmd=_login-run

Sincerely,
Paypal customer department

http://66.160.154.156/catalog/paypal/

Please do not reply to this e-mail. Mail sent to this address cannot be answered. For assistance, log in to your PayPal account and choose the "Help" link in the footer of any page.

To receive email notifications in plain text instead of HTML, update your preferences here.

How to drown in spam

Spam is junk mail. Unsolicited e-mail. It's among the most hated forms of advertising. About 70 percent of *all* e-mail on the Internet is spam.

There's no cure for spam once you start getting it. It is possible, though, to avoid getting into the spammers' address books in the first place.

Here's the main thing: Don't ever post your main e-mail address online—into a Web site that requests it, for example. Set up a different e-mail account and use *that* address for online shopping, Web site and software registration, and comment posting. Otherwise, your e-mail address will be sold, along with millions of others, on massive lists that wind up in the hands of spammers.

Bonus tip: You can sign up for a temporary, 10-minute e-mail address for use on Web store forms; see page 230.

Even then, whenever you fill out a form online, look for checkboxes requesting permission for the company to send you e-mail or to share your e-mail address with its "partners." Just say no.

If your spam problem is already out of control, and you don't want to start fresh with a new e-mail account, buy an antispam program like SpamAssassin (Windows) or SpamSieve (Mac). These programs aren't perfect, but they reduce the flood by quite a bit.

How to humiliate yourself in front of everyone

Every e-mail program has *two* buttons that let you reply to a message. One says Reply. The other says Reply All.

Reply sends your answer back to *one person:* whoever sent the message. Reply All sends your response to *everyone* who got the message in the first place. That could include a couple of other people. It could include dozens of other people. It could include your entire company.

Everyone—*everyone*—will, at some point in life, make the horrible mistake of clicking Reply All by accident.

Why is that horrible? Because often, you *intended* your reply only

for the person who wrote the original note. "OH, I can't stand that guy!" you might say, unaware that "that guy" is among the people getting your reply.

There's no Undo when you send an e-mail. If you click Reply All accidentally, only one feature can spare you from humiliation or firing: writing up a quick, heartfelt apology—and clicking Send.

How to let a virus take down your PC

If it weren't for the doggone Internet, computers would be a lot of fun.

But no, we've managed to interconnect all our computers via the Internet. We've given the bad guys a way to enter our machines, install viruses, set up remote hacking tools, feed us spyware, and otherwise turn our lives into an endless troubleshooting session.

But here's the surprise: *They can't do it without your help.*

They can't just *shove* a virus onto your PC. They can only set traps for you. *You* are the one who clicks the link, falls for the fake ad, or opens the file that winds up installing the spyware or the virus.

If you have a Windows PC, you should also have an antivirus program, like the free ones described on page 147.

But even then, you can take steps to avoid malware. (*Malware* is any kind of software that's designed to gum up or take over your PC.)

Often, the bad guys set their honeypot traps in exactly the kinds of places you might expect: Web sites that offer illegal access to pirated music, movies, and software. Porn sites. Ads on the Web, and in e-mail, that offer deals that are too good to be true.

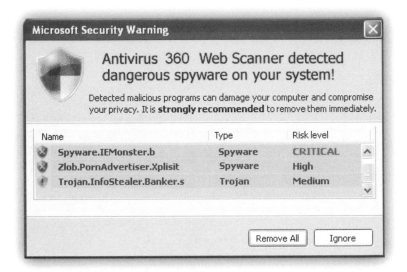

Sometimes, when you're on the Web, a "Virus found!" message appears. You're supposed to think that it's your antivirus software at work—but actually, it's a trick. If you click anything in the ad, you install the bad software behind the scenes.

With e-mail, don't open file attachments from somebody you don't know; e-mail attachments are a *very* common way of getting yourself infected.

Furthermore, don't open any attachment from someone you *do* know unless you are expecting it. (Your friends' computers might be infected by software that blasts out e-mail without their knowledge.)

A clear-headed lecture
about passwords

For years, the most commonly chosen password in the world was, believe it or not, the word *password*.

Fortunately, people are getting smarter. As of 2014, *password* was no longer the No. 1 most used password.

The new No. 1 password? *123456*. Good job, America.

The usual advice is "Make up a different, complicated password for every Web site—letters, numbers, punctuation, no recognizable words. And *change* every password every 30 days."

Yeah, sounds great. Except *nobody* would do that. Nobody *could* do that.

So here's some more reasonable advice.

- **Use a different password for every *important* site.** Your bank, your e-mail account, Facebook, Twitter, Amazon. If the villains manage to get *one* of your passwords, at least they won't be able to get into *all* of your accounts.

  It's not nearly as critical to use different passwords for all your *other* sites: news sites, sports sites, blogs, and so on. (What are the bad guys gonna do—see which sports teams you follow?)

- **Make up a password you can remember—using the first letters of a sentence.** For example, the password *iwih25md* looks impossible for a hacker to guess—and impossible for you to memorize. But actually, it stands for *I wish I had 25 million dollars.* You can remember *that,* can't you?

  Here's another trick for using a different password for every site without having to become a national memory champion: Change the password for each site by *one letter,* corresponding to the name of the Web site. For Twitter, "iwih25m**dt**,"; for Citibank, "iwih25m**dc**."

- **Install a password manager.** The *best* solution to the too-many-passwords problem, really, is to install a password-*management* program like Dashlane (free) or OnePass. Or turn on the iCloud Keychain feature for Apple phones, tablets, and computers. These programs fill in your passwords *automatically*. So you can have a different, complex password for every site you visit—without your having to memorize anything at all!

New Customer

Please specify your **contact** *information.*

Name:	
Phone Number:	
E-Mail Address:	
E-Mail Address: Re-type to confirm	

New Customer

Please specify your **contact** *information.*

Name:	Casey Robin
Phone Number:	516-512-3344
E-Mail Address:	caseyrobin@gmail.com
E-Mail Address: Re-type to confirm	caseyrobin@gmail.com

(These programs can also store your credit-card details, so you don't have to type all that stuff out every time you buy something.)

Set up an automatic
backup system—today

You want to hear an incredible statistic? Guess how many people have complete, up-to-date backups (safety copies) of everything on their computers.

Four percent.

The rest of us are just tempting fate. Every hard drive in every computer *will* someday fail if it runs long enough. Where will your photos, movies, music, documents, and software collection be then?

The reason that percentage is so low, of course, is that setting up a backup system isn't cheap or easy. It's a lot of steps, and nobody shows you how.

One approach: Pay a monthly fee for a service like Mozy or Carbonite. These companies offer you software that continuously backs up your computer *online.* That's smart, because your backup will be safe if your computer is stolen, flooded, or burned. But keep in mind that if the worst should come to pass, downloading your files again is a very slow process; it can take several days.

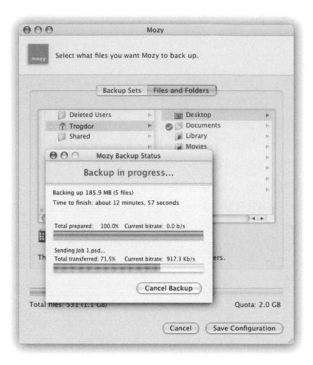

Another approach: Buy an external hard drive to hold your backed-up files. They're incredibly inexpensive; you can get a 2-terabyte drive (that's 2,000 gigabytes) for $75 these days.

Then you need some software that will automatically back up your files onto that drive. Read on.

Automatic backups on the Mac

When you connect a new hard drive to your Mac, a message asks if you want to use it for *Time Machine*, Apple's built-in backup software. Click "Use as Backup Disk." From now on, the Mac will back up your entire computer on that drive. (Actually, it can be another *internal* drive, or even another Mac on the network.)

Every hour, it stores on that drive only the files that have *changed;* it doesn't make another entire copy of everything.

In System Preferences (page 111), you can click the Time Machine icon to make changes, exclude things from the backup, and so on.

Now, then: Suppose the terrible day comes. You lose or delete a file, or want to recover an older draft of something. You're in luck.

Open the disk or folder window where the file was. Now choose Enter Time Machine from the ⏱ menu.

Your desktop slides down to reveal some classic Apple eye candy: an animated starry universe. The desktop window you opened seems to be one of hundreds, stretching back in time.

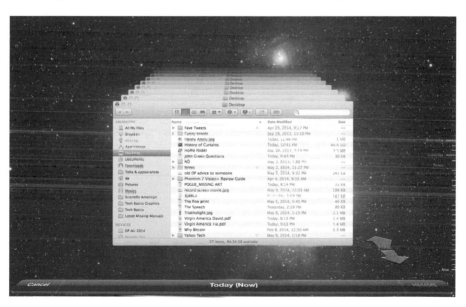

Each is a snapshot of that window at the time of a Time Machine backup.

You can drag your cursor through the timeline at the right side, as though it's a master dial that flies through the windows into past versions. You can use the search box in the corner of the window.

Or you can click one of the two big, flat perspective arrows. The one pointing upward means "Jump directly to the most recent window version that's different from the way it is right now."

Eventually, you'll find the deleted or changed file. Click it and then click Restore (lower right). The OS X desktop rises again from the bottom of the screen; there's a moment of copying; and then presto: The lost file or folder is back in the window where it belongs.

Automatic backups in Windows

If you're willing to invest in a backup hard drive, you'll really like File History, the Windows automatic backup system. If anything bad happens to the files you work on—including *you* making an ill-advised revision on too little sleep—you can rewind to a time when it was safe, and recover it.

When you connect an external drive, a Windows AutoPlay box offers to "Back up your files on this drive." That's what you want. The rest is automatic; Windows quietly backs up your PC once an hour, so it'll be ready in case of disaster.

When the day comes that you want to recover a lost, deleted, or badly edited file, open the Control Panel pane again. Click "Restore personal files."

Now you see this window:

Your job is to find the file you want to recover. You can dig it up manually, double-clicking folders as usual.

If it's been deleted, of course, you won't find it without rewinding its window into the past. You do that by clicking the (↺) button. The entire window slides to the right, showing you *versions* of the current window, going back in time. Scroll back far enough, and you'll eventually see the missing item reappear. (At any time, you can double-click a document to open it in this window, so you can see if it's the correct draft.)

You can also *search* for the missing item by typing its name into the search box at the top of the window.

Once you've found the file (or a good version of it) or folder, click the big green Restore button. Windows brings back the lost file, folder, or document from the dead.

Rewind your entire PC to a time when it worked better

When the day comes that your PC suddenly starts acting up, you can "rewind" it to an earlier condition when everything worked fine—without changing any of your e-mail, documents, and so on. This delicious feature is called System Restore.

Unbeknownst to you, the Windows System Restore feature has been creating memorized snapshots of your PC's copy of Windows, called *restore points,* ever since you've been running it. Once a day, plus every time you install anything new. When the day comes that your PC suddenly starts acting up, you can "rewind" it back to a restore point when everything worked fine—without changing any of your e-mail, documents, and so on.

If this is your unlucky day, here's how you do the rewinding. First, open the System Properties dialog box. (Here's one way to get there: Open the Control Panel. In its search box, type *properties;* in the results, click System.) At the left side, click System Protection.

Now click the System Restore button, as shown here.

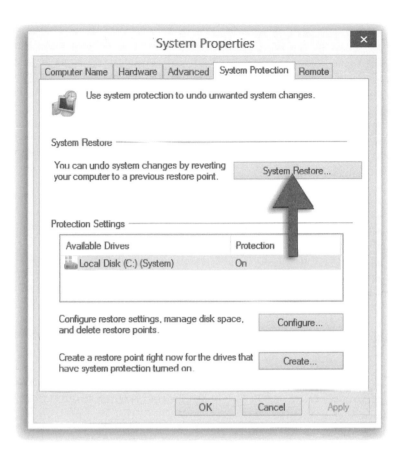

Next, click Next.

Now Windows displays a list of restore points; a little description helps you understand what changes you made to your PC on that day.

With a little bit of study, you can usually figure out the best one to rewind to. Click its name, as shown on the next page.

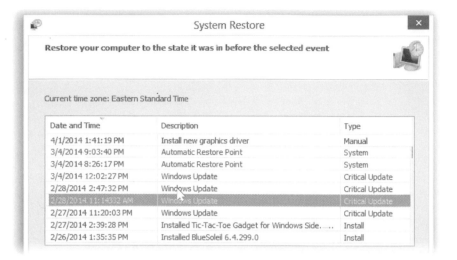

Date and Time	Description	Type
4/1/2014 1:41:19 PM	Install new graphics driver	Manual
3/4/2014 9:03:40 PM	Automatic Restore Point	System
3/4/2014 8:26:17 PM	Automatic Restore Point	System
3/4/2014 12:02:27 PM	Windows Update	Critical Update
2/28/2014 2:47:32 PM	Windows Update	Critical Update
2/28/2014 11:14:32 AM	Windows Update	Critical Update
2/27/2014 11:20:03 PM	Windows Update	Critical Update
2/27/2014 2:39:28 PM	Installed Tic-Tac-Toe Gadget for Windows Side.....	Install
2/26/2014 1:35:35 PM	Installed BlueSoleil 6.4.299.0	Install

Click Next again, then Finish, then Yes, then marvel as Windows reinstates your operating system to its condition on the date you specified. Don't use your PC during this process.

Once the PC restarts, you should be back in business. (If you don't like the results, you can *undo* the undoing. Just click Undo System Restore, which appears on the System Protection box shown above.)

Part 3

The Internet

Chapter 10: **E-mail**

A h, e-mail: Fount of a thousand wonders, bearer of joyous tidings, builder of careers.

And also the source of a million frustrations, setup for humiliations, ender of relationships.

There'd be a lot more of the good stuff, and less of the bad stuff, if everybody knew what you're about to know about the art and science of e-mail.

The 10-minute e-mail address

Often, when you sign up for a new Web site account, the site says: "We've sent an e-mail to the address you've supplied. Your account will not be active until you click the verification link in that message."

And sure enough: In a little while, a message like this appears in your e-mail:

Amazon.com   December 2 4:39 PM
To: david@pogueman.com   Hide Details
Reply-To: account-update@amazon.com   Inbox
Please verify your e-mail address for Amazon.com

Thank you for verifying your e-mail address. Please click the following link to complete the process:

https://www.amazon.com/ap/emailverify?aU=Ja1a0cf0 C4b1 1066 9195-03a13d011777

If you did not request to have your email verified you can safely ignore this email. Rest assured your customer account is safe.

The purpose of that exercise is to confirm that you are *you*—that you really did *want* to sign up. Without that verification, some prankster could sign you up for things—you know, the "Hot Fridge Repairman Photo of the Day" newsletter—without your knowledge.

Unfortunately, as you know (don't you?), it's foolish to supply your actual, primary e-mail address on the Web. Providing your *real* address is just asking to land on the mailing lists of spammers (see page 213).

Here's a handy workaround: Use a *temporary* e-mail address. Go to *www.10minute-mail.com*. There, staring you in the face, is an e-mail address that's been generated just for you, right now.

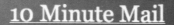

10 Minute Mail

≡ Welcome to 10 Minute Mail

Beat spam with the best disposable e-mail service.

a1257475@drdrb.net is your temporary e-mail address.

Click here to copy this e-mail address to your clipboard

a1257475@drdrb.net

Your e-mail address will expire in 6 minutes.
I need more time! Give me 10 more minutes!

Use *that* as your e-mail address. Any verification messages sent to that address appear *right here,* on this Web site. You can "click to verify" right there—and no spammer will ever get your real e-mail address.

How to confirm a juicy e-mail item before you pass it on

If you have an e-mail account, then you must also receive the occasional eyebrow-raising story, passed along to you by some relative or well-meaning friend.

The world is full of these stories: Your friend was mugged at an overseas airport and needs you to wire some money. Your cell phone number is now being sold to telemarketers. Obama's a Muslim.

Turns out those stories—and *most* of the others that get passed around by e-mail—are fake. They're great stories, all right. They go viral because they appeal to your sense of wonder, or outrage, or vengeance. But they're generally false.

There is, however, a wonderful place to find out before you become the sucker who passes one of those notes along. It's Snopes.com, a free site run by a husband-and-wife team in California. All they do, all day long, is research these things that get passed around—and report on their veracity. It's the world's clearinghouse for Internet scams and rumors.

Celling Your Soul

Claim: Cell phone users must register their numbers with the national "Do Not Call" directory by a given deadline to prevent their cell phone numbers from being released to telemarketers.

● FALSE

Origins: Despite dire warnings about the imminent release of cell phone numbers to telemarketers that continue to be circulated via e-mail year after year, cell phone users do not have to register their cell phone numbers with the national Do Not Call registry before a soon-to-pass deadline to head off an onslaught of telemarketing calls. The panic inducing e-mails (which circulate especially widely every January or June, since many versions of the warning list the end of those months as a cut-off date for

How to quote back (or forward) only an excerpt

When you want to respond to one point in someone's e-mail, drag through the text in question *before* you click Reply. Like this:

A quick question for you

Hi there... You may not remember me, but we met at the 2006 Drapery Distributors of America (DDA) southeast regional conference in South Carolina. Good times, right? Who can forget the rollout of DeSoto's new burnished burlap line? Now THOSE were some curtains!

Anyway, I write because you had mentioned reading a really good book that introduced kids to the whole field of drapery science. I've wracked both my brains and Amazon.com, and I haven't been able to find it. I'd be most grateful if you could remind me of the name of that book.

Meanwhile, I think of you often. Hope to see you at the next DDA meeting in Miami!

Casey Robin
Assistant to the Regional Sales Manager
Draperies of Detroit

Now, when you click Reply, your e-mail program gracefully "quotes back" *only* that portion of the original note. You can type your response directly beneath it, so that your correspondent knows what you're talking about.

| I'd be most grateful if you could remind me of the name of that book.

Why, yes—it was called "Curtains for Kids."

Hope this helps!

—Alex

Everybody's clear, and you look like a pro.

Send photos
that won't bounce back

It's a bummer, but it's true: *Full-size photos are too big to e-mail.*
Digital photos from a camera are enormous. They contain
millions of pixels (tiny color dots), because they're intended for
printing. And you need a lot more dots for a printout than you
do for looking at a picture on a *screen.* Like five times more.

Unfortunately, there's a size limit for e-mail attachments. It
depends on which e-mail service your recipient uses, but 10
megabytes is a typical limit. If you exceed that limit with your
attachments, your e-mail will bounce back to you, and your
intended recipient will never even know you made the attempt.

The world's software nerds have come up with all sorts of
different methods to help you get around this problem. Here
are a few:

- **Use the Mac's Mail program.** In OS X Yosemite, Apple
  has cleverly solved the attachment-size problem. If you send
  something big, your recipient sees a link that, when clicked,
  downloads your full-size originals. Even if they're 5 gigabytes big.

- **Use the E-mail command in the Pictures folder.** If your
  PC runs Windows 7 or Windows 8, life is easy. Click the
  photo you want to send. (To send several, Ctrl-click them.)
  Then, on the Windows 7 toolbar, click E-mail; in Windows
  8, open the Share menu and choose E-mail, as shown on the
  following page:

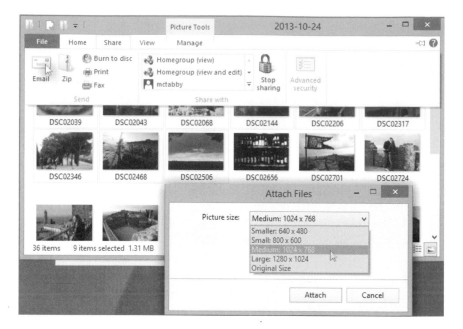

Either way, Windows now offers to shrink down your pictures to e-mailable size. From the pop-up menu, choose Medium or Large. The dialog box tells you how big the resulting package will be; note that 10 MB (megabytes) is usually the most you can attach to an e-mail message.

When you click Attach, Windows opens an outgoing e-mail message—with the scaled-down photos already attached.

- **Use your photo program's Send E-mail command.** If you use a program on your computer to manage your photos—like iPhoto (Mac) or Picasa (Mac or Windows)—you're all set. These programs offer E-mail buttons that shrink the photos and attach them to an outgoing e-mail message.

  In iPhoto, for example, select the photos, click the Share button, then click E-mail.

- **Put the photos into Dropbox.** If you've signed up for the free Dropbox service (page 105)—and you should—your problems are over. You can put your photos into a folder inside your Dropbox folder at *full size*—and then share that folder electronically with other people.

  This method sends them the full-size photos, suitable for printing or anything, by bypassing e-mail altogether.

- **Just post them on the Web.** Finally, consider *why* you're trying to e-mail photos. If it's so your mom or your kid or your friends can see them, wouldn't it be easier to post them in a gallery on the Web?

  If you have a Flickr.com account, for example, you can specify *exactly* who gets to see your photos, even if that's just one person; the same is true for Facebook.

  Posting your photos is a great way of creating online albums and controlling who gets to see them.

The peril and promise of BCC

When you address an e-mail message, you're offered a "To:" box, in which you enter an e-mail address. You're also offered a "Cc:" box, in which you can enter an e-mail address of someone *else* who might be interested (but who isn't the main recipient). It stands for "Carbon copy."

Your e-mail program also offers a third address box, labeled "Bcc." It stands for "*Blind* carbon copy." (In many e-mail programs, this box comes hidden; you have to click some button or setting to reveal it.)

Bcc lets you send a copy of a message to somebody *secretly;* none of the other recipients knows. Everyone sees the addresses in the "To:" and "Cc:" boxes, but nobody can see who's getting Bcc copies.

People send Bcc copies when they want to tip off a third party. For example, if you send your boss a message that says, "Hi boss—I've doctored our financial books according to your instructions," you could Bcc the FBI to clue it in without getting into trouble with your boss.

The Bcc also is useful when you want to send a message (like a joke) to a *lot* of people. If you put everyone's address into the Bcc box, no one will have to scroll through a long, ugly, privacy-invading list of e-mail addresses to get to the message part.

What *this* means online

Ever wonder why some people put asterisks around words online, *like this*?

In the olden days, people couldn't use italics in e-mail, *like this.* So in those dark days, *asterisks* were the e-mail version of *emphasis.*

Some people still use that notation out of habit. Some people use that notation because they send "plain-text" e-mail (no formatting allowed). And some people still use that notation on online services like Twitter and Facebook, because those still don't let you use bold and italics. (Same for phone text messages, by the way.)

The non-teenager's guide to texting shorthand

It all started with early cell phones. Back in those ancient days, typing was tedious and miserable; you had to tap out your messages on a *numbered* dialing pad. No wonder the world quickly adopted shorthand phrases.

The funny thing is, though, people still use them today. And not just on phones. These abbreviations crop up in e-mail messages, chat rooms, discussion sites, and—yes—even school writing assignments.

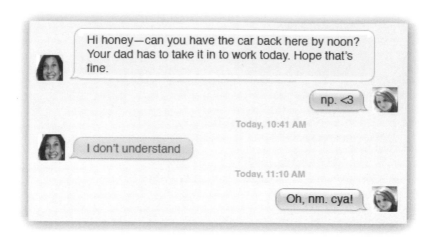

Nobody ever sits down and teaches you what they stand for, though—so let this page be your gentle tutor.

?	I don't understand what you're saying.
.02	That's my two cents' worth.
<3	I love you. (It's a heart sideways.)
brb	Be right back!
btw	By the way…
cya	See you!
ftw	For the win! (Meaning, "That's the BEST!")
fwiw	For what it's worth
gtg	Got to go
idk	I don't know.
iirc	If I recall correctly…
imho	In my humble opinion
irl	In real life
jk	Just kidding
kk	OK
lol	Laughing out loud
meh	I feel so-so about that
noob	Newbie (beginner)
np	No problem

nvm	Never mind.
otoh	On the other hand
rofl	Rolling on the floor laughing
rtfm	Read the [freaking] manual!
sup	What's up?
ttyl	Talk to you later
uok	Are you OK?
wrt	With regards to
wtf	What the [heck]!?
wtg	Way to go!
ymmv	Your mileage may vary.

Two quick ways of attaching a file to an outgoing message

When you want to attach a file to an e-mail you're sending, you might instinctively click the little paper-clip icon. But there might be an easier way.

You can drag a file—or even several—right off the desktop and into the outgoing e-mail *window*, like this:

That trick involves positioning your e-mail window so that you can *see* the desktop, of course.

If you're already *at* the desktop, another method awaits: the right-click method.

- **Windows:** Right-click the icon you want to attach. From the shortcut menu, choose "Send to"; from the shortcut menu, choose "E-mail recipient."

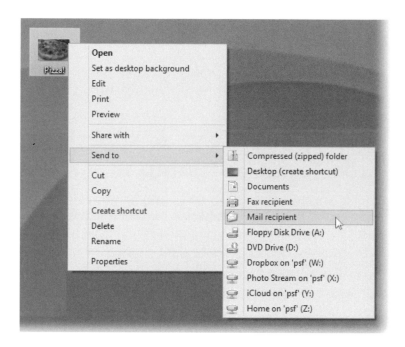

- **Mac:** Right-click the icon you want to attach (page 79). From the shortcut menu, choose Share; from the shortcut menu, choose E-mail, as shown on the facing page.

(These tricks work only if you do your e-mail in a *program,* like Outlook, Windows Live Mail, or Apple Mail. These shortcuts don't work if you do e-mail on a Web site, like Gmail, Yahoo Mail, Hotmail, or Outlook.com.)

How not to lose attached files

When someone sends you a file attached to an e-mail, you'll know it. The traditional paper-clip icon appears along with the file's name.

As you know from page 215, you should never open an attachment from a stranger, and never open one from someone you *do* know unless you are expecting it.

Here's something else to worry about, though: losing the attachment into the bowels of your computer's folder system.

You can double-click the file's icon to open it, look at it, and even edit it. But at that point, the real file is still embedded in your behind-the-scenes e-mail stash. It hasn't been freed and released into the visible realm of your own desktop and folders. To do that, you must do one of these things:

- Double-click the attachment's name or icon. It opens right up in Word, Excel, or whatever. Now click the File menu and choose Save As; choose a location for it.

- Drag the file's icon *out* of the e-mail window (or Web-browser window) and onto any visible part of your desktop behind it.

- If you use an e-mail *program,* like Outlook, Windows Live Mail, or Apple Mail, you can right-click the attachment's name and, from the shortcut menu, choose "Save as." (In Apple Mail, it's called Save to Downloads Folder.)

In Mail on the Mac, you may find it easier to click the little attachment menu shown here—and choose Save All.

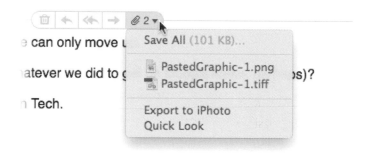

Any of these options takes the file out of the Mail world and into your standard Windows or Mac world, where you can file it, trash it, open it, or manipulate it as you would any file.

Chapter 11:
Web Browsers

When you let someone loose on the World Wide Web (as your ancestors used to call it), you're really expecting that person to learn two things simultaneously.

There's the Web itself: that infinitely vast, ever-changing, endlessly interlinked network of pages; the world's biggest source of information, entertainment, titillation, and shopping.

And then there's the program you use to *navigate* the Web: the browser.

It's a program so essential, it comes built into every computer, tablet, and smartphone ever made   and yet it comes without so much as a Quick Start Guide.

Here are the basics you've been missing.

The mighty ".com" shortcut

When you surf the Web, one thing you probably do a lot is type *Web addresses* into the bar at the top: *nytimes. com, youtube.com, iwastesomuchtime.com,* and so on.

Logic should tell you, therefore, that the less of that typing you have to do, the better and longer your life will seem.

That's why you should learn the Mighty .Com Shortcut.

When typing a Web address into your Web browser, you can leave off the "http://" part. Just type the rest: *nytimes.com* or *youtube.com,* for example. Your Web browser is smart enough to know what you mean.

In fact, for *most* sites, you can also leave off the "www." part. (It varies by Web site.)

You don't have to type ".com" at the end, either:

- **Chrome, Internet Explorer, Firefox:** After you've typed the main part of the address (like *nytimes* or *amazon*), just press Ctrl+Enter. Your browser automatically adds ".com" and takes you to the site!

  (On the Mac version of Firefox, it's ⌘-Enter instead.)

  Bonus tip: If you press *Shift*+Enter, Internet Explorer adds ".net." If you press Shift+Ctrl+Enter, you get ".org."

- **Safari:** Alas, there's no way to make Safari add the *.com* automatically. (You can leave off the *http://* and *www.* portions, though.

Go back faster

Surfing the Web involves visiting one page after another—and often going *back* to a previous page. That's why there's a prominent Back button at the top-left corner of every Web browser—and why every Web browser offers a teeming multitude of *shortcuts* for going Back.

- **Backspace.** How awesome is this? Just tapping your Backspace key (on the Mac, called Delete) takes you *back* a page. Works in every browser, in Mac and Windows, except Safari.

- **Alt+←.** Yes, that's right: On Windows, it's the old Alt-left-arrow-key trick. It means "Back." (Once you've done that, you may also enjoy pressing Alt+→. It means, of course, "Forward again.")

  In Mac browsers (Safari, Chrome, Firefox), press ⌘-← instead.

- **Swipe two fingers.** In Safari or Chrome on the Mac, and if you have a trackpad: Swipe *two fingers* leftward across the trackpad. Cool!

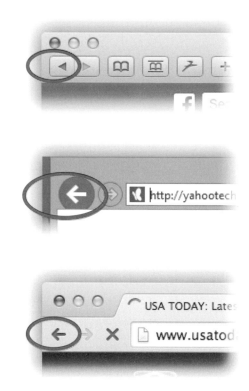

A faster way of getting to the Address bar

The Address bar is that all-important white strip at the top, where you type the address of the site you want to visit.

You can, of course, click there with the mouse, delete what's there, and type a new address. And that's great—if you don't think your hand gets enough exercise.

But there are faster, better ways.

- **Windows:** Press Alt+D.

- **Mac:** Press ⌘-L.

In both cases, you've just highlighted all of the text in the Address bar. Don't bother pressing Backspace or Delete—just *start typing* the new address. Your hands never leave the keyboard.

Bonus tip: If you must use the mouse or trackpad, you don't have to run your cursor across the address that's already in the box. Instead, click once on the *Web site logo* at the left side of the bar, as shown on the next page.

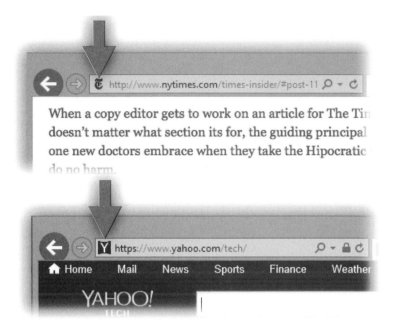

Doing that selects the existing address so that you can type right over it.

Print or e-mail the article without the ads

Many Web sites offer a Print button. If you see it, use it; it produces a version of the same Web page *without any ads or clutter.* Just the text, as shown on the facing page at right.

That's great for printing, of course—but it's also great for e-mailing to a friend. Press Ctrl+A to select all of the text, then Ctrl+C to copy it to your clipboard; switch to your outgoing e-mail message, and finally press Ctrl+V to paste it. (On the Mac, use the ⌘ key for those keystrokes instead of Ctrl.)

Pure, clean, easy to read—and entirely free of ads.

Shorten and simplify a Web address

There's no limit to the length of a Web address—unfortunately. Even a Web page that's only a paragraph long might be waiting for you at:

http://developers.jollypad.com/fb/index. php?dmmy=1&fb_sig_in_iframe=1&fb_sig_iframe_key=8 e296a067a37563370ded05f5a3bf3ec&fb_sig_locale=bg_ BG&=1282749119.128&file_update_time=expire55600&fb_ sig_user28&fb_sig_session_key=2.IuyNqrcLQaqPhjzh_. ea9498aabcd1b_sig_app_id=177509235268&fb_sig=1a5c6100f a19c1c9b983e2d6ccfc05ef.html.

There are times, though, when a long URL (Web address) isn't what you want—for example, when someone has to type it out from a printed page. Or when a huge blob like that would seem intimidating or too technical—in an ad, a poster, a homework assignment. Or anywhere with limited space.

That's why it's worth knowing about URL-*shortening* sites like bitly.com. All you do is paste your long address into the box at the top of the page ...

... and click Shorten. Presto: You get a *very* short address that takes you to the exact same page. Click the Copy button to copy it to your clipboard, ready to paste wherever you need it.

It's free, it's instantaneous, and you don't have to sign up for anything.

A crash course in tabbed browsing

T abbed browsing is a way of keeping a bunch of Web pages open simultaneously—in a single, neat window, like this:

Hardcore Web surfers adore this feature. It gives them a sense of control and order.

To create a new tab, click the stubby blank "file-folder tab" at the top right of your window, as shown on the next page.

Or press the magic keystroke:

- **Mac browsers:** ⌘-T (Safari, Chrome, Firefox)

- **Windows browsers:** Ctrl+T

Or, if you're about to click a link on a Web page before you, you can open it into a new tab directly by pressing that same key as you click: Ctrl on Windows, ⌘ on Mac.

Chrome

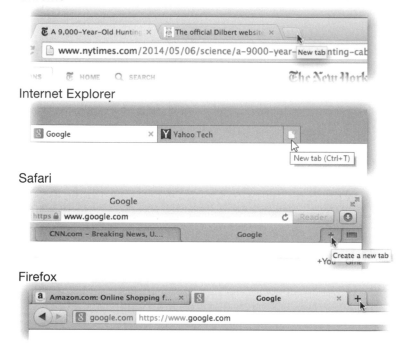

Internet Explorer

Safari

Firefox

Once some tabs are open, you can switch among them from the keyboard, too. On Windows, press Ctrl+1, Ctrl+2, Ctrl+3, and so on (to flip to the first, second, and third tabs). On the Mac, it's ⌘-1, ⌘-2, ⌘-3, and so on.

(That doesn't work in Safari.)

Finally, you can press Ctrl-Tab (on the Mac, Control-Tab) to rotate through your open tabs. Even in Safari.

Add the Shift key to move *backward* through your open tabs.

The secret home-page logo

On commercial Web sites like Yahoo, YouTube, Facebook, Amazon, NYTimes.com, and so on, the logo in the upper-left corner is actually a button. You can click it to return to the site's home page.

That's handy when you've drilled down to some article and want to jump back to your starting place.

The Space bar scrolls down one screenful

O ccasionally you'll find an article on the Web that's very short. It fits completely on your screen without scrolling.

But most of the time, you have to *scroll* to read the whole article.

You can certainly use the mouse to scroll down, using the scroll bar at the right side of the window, just as you'd scroll down in any other program.

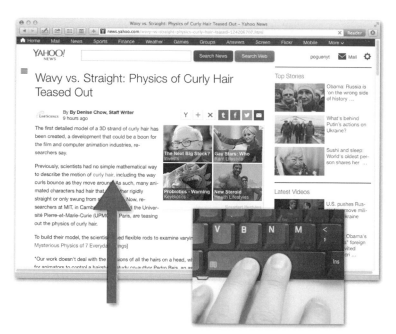

But for a task that you wind up performing many times a day, that's a lot of manual, fussy effort. Unless you're paid by the hour, there's a much better way:

Tap the Space bar.

Yes, the Space bar means "Scroll down by exactly one screenful." It works in every Web browser, on every kind of computer, and it's well worth committing to your muscle memory.

And once you've mastered that step, here's a bonus trick: If you're pressing the Shift key, tapping the Space bar makes the page scroll back up again.

The Tab key moves from box to box

When you're filling in a form online, you can press the Tab key to jump from box to box. You can—and you should.

Why is that better than using the mouse to click inside each box before you type? Because this way, your hands never have to leave the keyboard. It's much more efficient.

This Tab-key business works on Web pages. But it also works anywhere on your computer—including dialog boxes like the one shown on the next page.

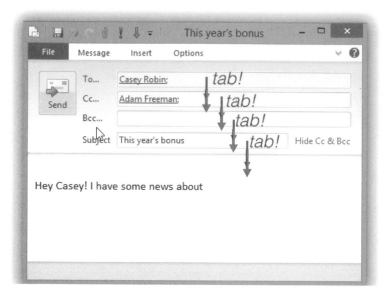

Bonus tip: Hold down the Shift key to jump *backward* through the boxes with each Tab press.

Bonus bonus tip: If you tab your way onto a *checkbox*, you still don't have to take your hands off the keyboard. Once the checkbox is highlighted, just tap the Space bar to turn it on or off.

Pop-up menus online: Type the first letter

When you're asked to choose from a pop-up menu online, you can type the first letter of the choice you want.

That's especially handy when there are a *lot* of choices—like when you're asked to choose your *state* from the list of 50, or your country from the list of 220! Using the mouse, you'd spend a whole weekend trying to find the one you want.

Instead, once you've highlighted the pop-up menu (by pressing the Tab key, of course), just type the first letter of your state or country.

Type that letter key repeatedly to cycle through the different state names that begin with that letter.

For example, to choose Texas from a "State" pop-up menu, press the letter T key twice. For California, just press the letter C key once.

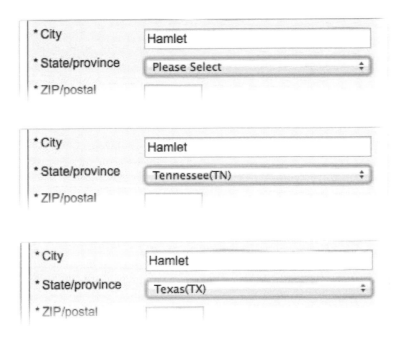

How to find a word
on a page

You probably know already how to find a Web page on the Internet: Type what you're looking for into the Address bar. (In most browsers, the Address bar is *also* the Search box. In a few older ones, the Search box is separate.)

But not everyone realizes you can also find a word *on a Web page* once you're there. That can save you a lot of reading and scrolling.

In most browsers, you just open the Edit menu and choose Find. (Sometimes you then have to choose Find again from the submenu.)

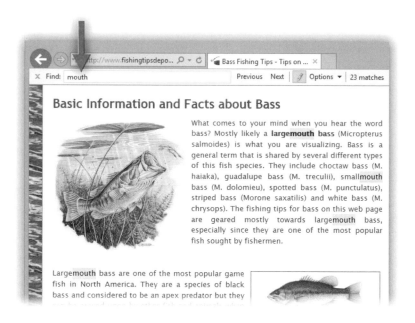

Of course, it's faster and easier to learn the keyboard short-cut: ⌘-F on the Mac, Ctrl+F on Windows.

And if the first occurrence of the word isn't the one you wanted, you can find the *next* occurrence by pressing—what else?—the *next* letter of the alphabet. That's ⌘-G on the Mac, Ctrl+G on Windows.

Print just what you want

Life would be glorious if every Web page showed *only* the stuff you wanted to read. But these days, the actual article is generally surrounded by a blinky Times Square of ads, tables, cross-references, "you may also enjoy…" links, and so on.

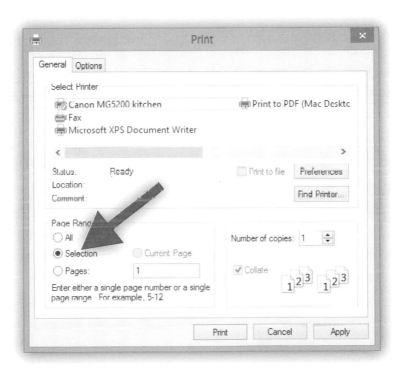

When you want to print the page, you'd be wise to enter the Printing mode first, as described on page 248. But you can do even better than that.

If you *drag through* some text before you choose the Print command, you'll print *only that*. Just remember to turn on the Selection box shown on the next page before you click Print.

This checkbox is available in all browsers except Safari. It's always hiding in the Print dialog box somewhere. In Firefox on the Mac, for example, you have to click Show Details to expand the options to see it.

You'll save all kinds of ink, paper, and headache.

How to eliminate ads and blinky things

I f you're like most other people, you're finding it increasingly difficult to enjoy the Web. It's tough to concentrate on what you're reading when you're bombarded by animation, blinking, and obnoxious ads blocking your view.

You should know, therefore, about free programs like Adblock Plus (for Windows, *adblockplus.org*) and AdBlock (for Mac, *getadblock.com*). They do exactly what it sounds like: They remove all the ads from your Web pages!

Now, advertising pays for a lot of the free material online, of course. So don't forget to weigh the moral implications of the act you're about to commit.

Enlarge that tiny type

Not every Web designer's eyes are the same age as yours, and not every Web designer's *screen* has the same resolution as yours. No wonder Web sites often seem to feature a text size best suited for gnats.

Fortunately, every Web browser makes it easy to enlarge the type on the page you're reading at that moment. All you have to do is press Ctrl+plus sign. Press it repeatedly, if you like, until the text is big enough.

That is, while you're pressing the Ctrl key, tap the + key as necessary to enlarge the text. Use the – key instead (also with Ctrl) to make the text smaller again.

On the Mac, use the ⌘ key instead. That is, press ⌘-plus to make the text larger, or ⌘-minus to make it smaller.

Bonus tip for Safari: If you open the View menu and choose Zoom Text Only, that keystroke will magnify or shrink *only the text* on your page. Graphics will remain at their original size.

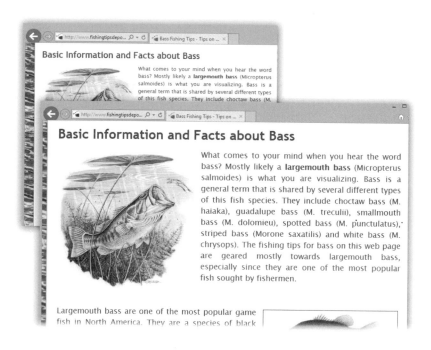

Privacy mode

Every Web browser today keeps track of which sites you open. Each Web page winds up listed in your browser's History menu, for example, to make it easy for you to find it again later. Your browser also downloads and saves *pieces* of the Web page (like graphics), which makes them load more quickly if you revisit that page.

There are times, however, when you might *not* want to leave a record of your online activities. For example, when you're shopping for gifts for your spouse. Or when you're using a public computer at a library, and you don't want it to store your account information. Or when you're visiting Web sites that, ahem, your boss or family might not approve of.

(It's even said that some airlines keep track of your flight searches—and if you book the same one often, the airline raises the price it shows you!)

But every browser today *also* offers a Privacy mode. In this mode, the browser doesn't maintain *any* record of your surfing activities. It doesn't add any pages to your History list, searches to your Google search suggestions, passwords to your saved password list, or autofill entries to Safari's memory. And it doesn't save any cookies.

The trick is to turn on this mode *before* you start browsing.

- **Safari:** From the Safari menu, choose Private Browsing. (If you have OS X "Yosemite," things work a little differently. From the File menu, choose New Private Window.)

- **Chrome:** From the File menu, choose New Incognito Window.

- **Firefox:** From the Tools menu, choose Start Private Browsing.

- **Internet Explorer:** From the A menu, choose Safety, and then InPrivate Browsing (or press Ctrl+Shift+P).

Once you turn off this mode (or close the "private" window), the browser once again begins recording the pages you visit— but it never remembers the earlier ones.

What happens in Private Browsing stays in Private Browsing.

Chapter 12: **Google**

Google's corporate goal is to amass all human knowledge in one place, and make it instantly available to everyone. It thinks that job will probably take 300 years.

Maybe so, but Google is certainly well on its way. Most people think of Google as a search site, a card catalog for the Internet, but that's only half the story. Less, really.

You might be astonished to find out what else Google is good for.

Google is a dictionary

Next time you're about to be trounced in Scrabble by someone who insists that *wagyu* is a real word, no need to hunt around for a physical dictionary. Just type *define: wagyu* into Google's search box, and boom: There's your distressing answer.

(Hint: It's a cow.)

Google is a translator

At translate.google.com, you can choose languages you want to translate from and to: from Italian to English, let's say.

Now you can paste in some text you've copied from an e-mail or a Web site. (Or you can paste in the address of a Web site.) In a flash, Google translates the text for you.

The translation is performed by software, so it's often a little rough. But it's usually enough to give you the gist—and hey, it's free.

Google has pictures of everything

Guaranteed: You'll be astonished at how often Google can find a picture of *exactly* what you're looking for.

You read an article about George Clooney's new romantic interest. Your recipe calls for a starfruit. Your kid is doing a school report about knights, and needs a picture of a *hauberk*. Wouldn't it be great if you could instantly summon a photo—or thousands—of that person, that fruit, or that armor?

Go to *images.google.com*. (Or just go to *Google.com* and click the Images button.) Type in what you're looking for. Press Enter.

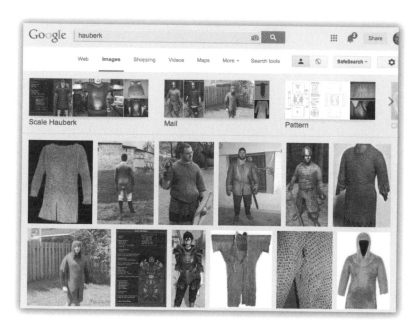

Isn't that amazing?

- -

Google knows what's in your picture

This one'll blow your mind.

If you offer Google a photo from your computer, it can find other photos like it. It somehow knows what's *in* the picture!

Try it. Go to *images.google.com,* which is Google's picture-searching site.

Click the little camera icon in the search box; you're offered the chance to choose a graphics file on your machine, or to paste in the Web address for a picture you found online. Or drag the icon of a picture off of your desktop into the Google search box, and press Enter.

In the search results, Google displays other photos that resemble yours. For example, if you took a picture of the Washington monument, you'll get thousands of other photos of the same thing.

Google also lists Web pages that *contain* those pictures. That's handy when you're trying to figure out what something is a picture *of*.

(This trick works well for pictures that might reasonably *be* on other Web pages—not so much for one-of-a-kind photos like your face scrunched up to a poodle's.)

Google is a travel agent

Next time you're looking for flights, Google is cleared for takeoff.

In the search box, you can type *new york London flights* (or which two cities you want). And boom: Google's first result is a tidy table of representative flights, listed by price.

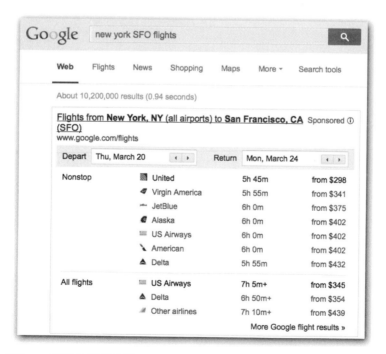

You can adjust the dates, using the controls above the table.

Click the title of the table, or any flight in the table, to view *all* the flights—and from there, you'll find links to Expedia, Travelocity, and the other online travel-booking services.

Google is an airport monitor

Next time you're supposed to meet somebody's flight, let Google tell you if it's going to be on time—and what terminal and gate it will use.

Just type *delta 2442* or whichever airline and flight number you want. When you press Enter, you get a dossier of details about the flight in progress: when it took off and from which gate, how much it's delayed, when it's arriving, and at which terminal and gate. It's great.

Google converts units of measurement and currency

Next time you have to convert anything to anything—yards to meters, grams to pounds, centigrade to Fahrenheit, euros to dollars—your first stop should be Google.

In that search box, you can type phrases like "teaspoons in 1.3 cups," or "miles in 2000 kilometers," or "32 C in F" (for temperatures), or "17 euros in dollars."

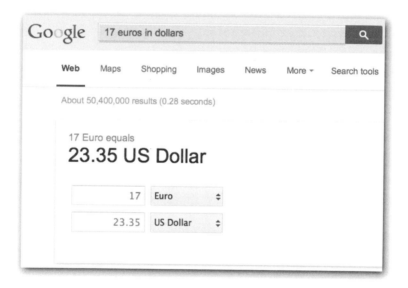

The wording in each case doesn't matter: You can type either "17 euros in dollars" or "dollars in 17 euros," for example.

Click Search (or press the Enter key) to see the answer.

Google is Mr. Moviefone

This one's the *greatest*.

In Google's search box, type *movies 10024* (or whatever you zip code is). You don't need to type the zip code if Google knows where you are; try it.

When you press Enter, Google shows you a tidy list of the most popular current movies, complete with their running times, ratings, genres, and a button that plays the trailer (ad) for each. Click "+Show more movies" to see a more complete list of current movies.

Movies for New York, NY				
Non-Stop	1hr 50min	PG-13	Action	Trailer
The Lego Movie	1hr 40min	PG	Animation	Trailer
RoboCop	1hr 48min	PG-13	Scifi	Trailer
About Last Night	1hr 40min	R	Comedy	Trailer
The Lego Movie in 3D	1hr 40min	PG	Animation	Trailer
Frozen Sing Along	1hr 48min	PG	Animation	Trailer
+ Show more movies				

To see where one of these movies is showing, click it. You get a magnificent listing of all local theaters where this movie's playing, complete with the showtimes.

The most information-dense movie page of all appears when you click "Movies for [name of the city]," shown in this illustration by the cursor. It presents a massive listing of every current movie and which theaters are showing it at which times, with links to their trailers, still photos, and IMDB pages. (That's the Internet Movie Database, a Web site dedicated to the details, cast lists, trivia, and viewer ratings for every movie ever.)

Non-Stop

Trailer - Photos - IMDb 8+1 +55 Recommend this on Google

1hr 50min - Rated PG-13 - Action/Adventure/Suspense/Thriller
Director: Jaume Collet-Serra - Cast: Liam Neeson, Julianne Moore, Scoot McNairy, Michelle Dockery, Corey Stoll

Global action star Liam Neeson stars in Non-Stop, a suspense thriller played out at 40,000 feet in the air. During a transatlantic flight from New York City to London, U.S. Air Marshal Bill Marks receives a series of cryptic text messages demanding that he instruct the airline to transfer $150 million more »

Regal Battery Park Stadium 11
102 North End Avenue, New York, NY
2:10 4:50 7:30 10:10pm

Regal Union Square Stadium 14
850 Broadway, New York, NY
11:00am 12:00 1:50 2:50 4:30 5:30 7:20 8:20 9:25 10:00 11:00pm

UA Court Street Stadium 12 & RPX
106 Court Street, Brooklyn, NY
12:20 2:00 3:00 4:40 5:40 7:20 8:20 10:00 11:00pm

Bow Tie Chelsea Cinemas
260 West 23rd Street, New York, NY
11:45am 2:15 5:00 7:30 10:00pm

AMC Loews Newport Centre 11
30-300 Mall Drive West, Jersey City, NJ
12:15 2:30 5:00 7:15 9:45pm

AMC Loews 34th Street 14
312 W. 34th St., New York, NY
10:40 11:40am 1:30 2:30 4:15 5:15 7:00 8:00 9:55 10:45pm

Show more theaters »

Son of God

Trailer - Photos - IMDb 8+1 +133 Recommend this on Google

2hr 18min - Rated PG-13 - Drama
Director: Christopher Spencer - Cast: Diogo Morgado, Roma Downey, Amber Rose Revah, Andrew Brooke, Louise Delamere

This major motion picture event is an experience created to be shared among families and communities across the U.S. It brings the story of Jesus' life to audiences through compelling cinematic storytelling that is both powerful and inspirational. Told with the scope and scale of an action more »

Google is a weatherman and an almanac

In Google's all-knowing Search box, you can type *weather new orleans, unemployment london,* or *population belgium.*

When you press Enter, Google instantly shows you the weather report, unemployment graph, or population graph for that town, state, or country.

Google is a package tracker
—and a flight tracker

Paste or type a package tracking number (from Fedex, UPS, or the U.S. Postal System) into Google's search box, and boom: Google shows you exactly where your package is.

Or paste the tail number of an airplane you spot, either on the ground or in the air, like N707JT—and Google shows you photos of that plane, its registration information, plus links to pages that describe it in detail.

Images for **N707JT** - Report images

N707JT Aircraft Registration FlightAware
https://flightaware.com/resources/registration/**N707JT** ▾ FlightAware ▾
N707JT FAA Registration information with aircraft photos, flight tracking, and maps.

N707JT FlightAware
https://flightaware.com/live/flight/**N707JT** ▾ FlightAware ▾
N707JT Flight Tracker (en route flights, arrivals, departures, history) with live maps and aircraft photos.

John Travolta's Private Boeing 707-138B [N707JT] Takeoff From Los ...
www.youtube.com/watch?v... ▾ YouTube ▾
Mar 1, 2012 - Uploaded by SpeedbirdHD .
Manufactured in 1964, this bird has been around...ex-Qantas,
 ► 0:52 ex-Braniff, now sporting the retro Qantas livery ...

Google is a calculator

Google's Search box can also do math for you. Type whatever equation you want in there—for example, 28*93+234/3=.

(On computers, phones, and tablets, the * means "times," and the / means "divided by.")

When you press Enter, Google shows you the answer—and presents a full-blown, on-screen calculator, complete with logarithmic functions, constants (like e and pi), and functions like Cos and Sin.

But that's just the beginning. Google's calculator is practically a math major. You can get the answer to math expressions like "6 factorial," or "cube root 6," or "pi^2", or "log 10 * ln e", or "arctangent 3."

Google's calculator can even translate numbers into binary code, which should cause programmers worldwide to celebrate. Just type "in binary" after the equation.

Google is a time-zone calculator

Want to know what time it is right now in Barcelona? Then go to Google and type in (what else?) "time in Barcelona."

(Handy hint: It works for other city, state, and country names, too.)

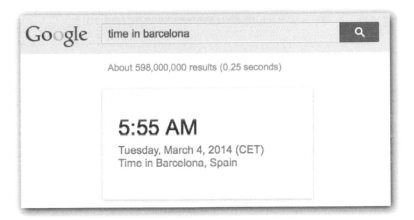

Google is a comedian

Go to Google and search for the word *askew*. Let's just say that Google doesn't just show you links to what the word *askew* means—it *illustrates* it.

Google is a Yellow Pages

You don't need a phone book anymore. The Google business directory is much more complete and up to date than the old Yellow Pages that used to wind up on your doorstep. Oh, and it covers the *entire world*.

Just type "windshield repair Miami," or "drugstores 44120,"

or "Thai restaurants Denver," or whatever. Or the name of the place you want: "H&R Block Chicago."

And boom: There is *everything* you might want to know. A map, showing where all the matches are. Phone numbers, addresses, links to the Web sites—all right there on the Results page.

Google knows the neighborhood

Google's thousands of computer-science nerds don't just sit there holed up in Silicon Valley all year long. They do get out some.

Since 2007, Google has sent fleets of Street View cars driving the roads of the world, *taking pictures* of what's on them. So far, it has logged 5 million miles of roads in 3,000 cities in 40 countries.

And why? So that you can type in a street address—and see what that building *looks* like.

That's incredibly useful when you're supposed to meet some-one at a new restaurant, or when you want to scope out the neighborhood where your job is sending you, or when you're shopping for a new house and want to "look around" before you pay a visit, or when you want to see what your old neighborhood looks like now.

Or when you just want to pretend you have a teleporter.

On your computer, go to *maps.google.com.* Type in the address you want. (Google is *very* forgiving of spellings and abbreviations. In fact, it'll probably propose the complete address when you've typed only a few letters.)

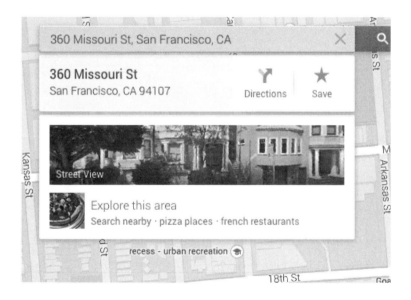

Once it has found the address, just click the Street View photo (visible in the illustration above)—and then marvel as the screen changes.

This isn't a *photo*, frozen in time. You can actually drag your mouse across it, up or down, to "look around" in all different directions. And you can click a piece of any road to "drive" in that direction. You could, if you really felt like it, click your way across the entire United States just like that.

Google Is Caesar

Google converts Arabic numerals to and from Roman numerals.

If you type *1587 in roman* into the search box and press Enter, Google shows you its Roman-numeral equivalent:

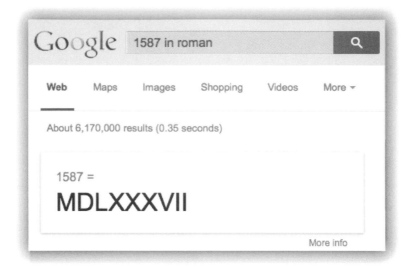

It works the other way, too. Type MCXCVII and press Enter, and Google reveals that that works out to 1197.

Google also does searches

Yes, OK, Google is a map, a phone book, a calculator, a translator, and a dessert topping. But get this: It also searches the Web.

You probably know the basics: Type something into the Search box and press Enter to see a list of Web pages that Google thinks might be what you're looking for.

Everyone and her brother knows *that* much. The true Google aficionado also knows this:

- **I'm feeling lucky.** If you click the "I'm feeling lucky" button, Google doesn't show you the usual list of search results; it takes you directly to the *first* item of the results.

  For example, if you're looking for the Web site of the Washington Redskins, just type *Redskins* and then click "I'm feeling lucky." You will go directly to the Redskins Web site (because it would have been the top search result).

- **Search for phrases.** If you put quotes around words in your search, Google finds only pages that contains those words *together*, in that order.

  For example, if you search for *hybrid helicopters,* Google shows you a list of all Web pages containing those two words—but not necessarily together. But if you search for

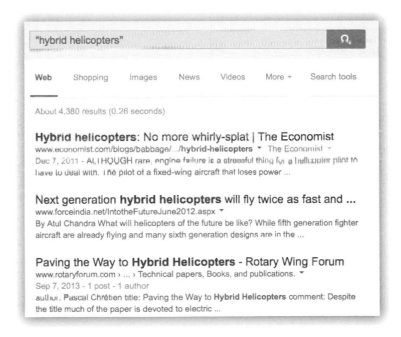

"hybrid helicopters," Google shows you pages that contain only that phrase—those two words together.

- **Skip the little words.** Don't bother typing little words like *the, of, and,* and so on. Google ignores them anyway; may as well save the typing.

- **Screen out the obvious red herrings.** If you put a minus sign in front of a word, you're saying, *"Don't* include any sites with *that* word."

Suppose, for example, that you search for *dolphins.* Half of your results will be Web pages about the *Miami* Dolphins team. But if you search for *dolphins–miami,* Google helpfully *leaves out* all Web sites about the team. It shows you only sites about the animal.

Chapter 13:
Videos Online

When people think of videos online, they usually think of
YouTube, of course—the biggest collection of videos in
human history. The numbers are staggering: 100 hours' worth
of videos is uploaded *every minute* of every day. And we, the
citizens of Earth, watch 6 billion hours of YouTube videos every
month. No wonder we get so little done.

This chapter covers video in all of its online forms: YouTube,
the Mother of All Video Sites, as well as the general concept of
watching TV shows and movies online.

The mystery of the YouTube Space bar

The main keystroke you need to know for YouTube is the same one you need to know for any video or music playback: the Space bar. Press it once to start playing a video, and again to pause.

"But wait!" protests the alert reader. "On page 254, I distinctly remember you saying that pressing the Space bar on a Web page *scrolls down!* So which is it, huh? On YouTube, does the Space bar start and stop playback? Or does it scroll down?"

Click here: Space bar plays

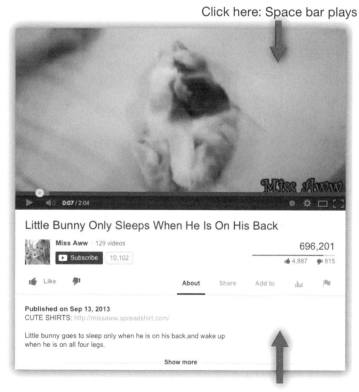

Little Bunny Only Sleeps When He Is On His Back

Miss Aww · 129 videos

Subscribe 10,102

696,201

4,887   815

Like

About   Share   Add to

Published on Sep 13, 2013
CUTE SHIRTS: http://missaww.spreadshirt.com/

Little bunny goes to sleep only when he is on his back,and wake up when he is on all four legs.

Show more

Click here: Space bar scrolls

You are absolutely correct, alert reader; there's a clash here. There's also an answer: When you first arrive at a YouTube video, tapping the Space bar *scrolls down,* so that you can read the thoughtful, measured comments left by viewers.

But if you *click the video first,* then Space bar plays and pauses, as you'd expect.

(Click *outside* the video, on a blank area of the page, and it's back to scrolling.)

The nearly unknown keyboard controls of YouTube

The Space bar isn't the only keyboard shortcut available to you, either. Check it out:

J Jump back 10 seconds. Great if you didn't catch the words.
K Play/Pause. (In case your Space bar is broken.)
L Jump forward 10 seconds. Great if the introduction is taking forever.
M Mute the audio.

(Isn't it neat how those four keys are all clustered together under your right hand?)

←, → Rewind, fast forward.
Home Jumps to the beginning.
End Jumps to the end.

number key:　0　　1　　2　　3　　4　　5　　6　　7　　8　　9

The number keys work, too. You can press 0, 1, 2, 3, and so on to jump to points in the video at those percentages in. 0 means "the beginning." 1 means "10 percent of the way through." 2 means "20 percent," and so on up to 9.

- -

Make YouTube videos bigger

YouTube videos start out filling only a small portion of the screen. The rest of the window is filled with other important elements, like ads.

But your computer screen is a glorious, colorful canvas, and you should *use* it! You should make the video fill your screen! And you can—in any of three ways:

- Click the small rectangle to make the video area twice as big.

- Click the lower-right control to make the video fill your screen *completely*. (Shrink it back down again by pressing the Esc key.)

Full Screen

Double Screen

- You can also go full-screen by double-clicking within the actual video image.

Make YouTube videos sharper

There are blurry YouTube videos, there are sharp YouTube videos—and there are blurry ones that you can *make sharper*.

Most of the time, YouTube's overlords (Google) send you a low-resolution, slightly "soft" (blurry) version of each video. Why? Because lower-quality (lower-resolution) videos play more smoothly on slow Internet connections.

So especially if you've enlarged your movie screen as described in the previous tip, you should definitely change the quality. You may be amazed at the difference.

To do that, click the ✿ icon shown here. A tiny panel shows you the current quality level. It's usually "Auto," which usually means "blurry."

But if you have a *fast* Internet connection, you may as well enjoy the higher resolution that most videos offer. Click the Quality pop-up menu and choose a highest number. On recent videos, you're likely to see "720p" or even "1080p," which are both hi-def quality. (The number indicates how many very fine lines make up the video picture. The higher, the better.)

If you watch a lot of YouTube, you may find that changing that setting every single time gets *old*. You may wish that You-Tube could do it automatically.

It can, actually. Look for another ✿ icon, this one in the upper-right corner of the window next to your name. (This icon appears only if you've signed in with a Google account or You-Tube account; if you haven't, click the Sign In button at top right.)

Click the ✿ icon; from the pop-up menu, choose YouTube Settings. At the left side of the screen, click Playback. Finally, turn on "Always play HD on fullscreen (when available)."

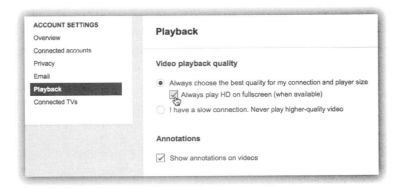

(While you're here, you may also want to turn off "Show annotations," so that your videos won't always be covered up with annoying little pop-up notes.)

Click Save, and enjoy your higher-quality life.

How to find YouTube videos more quickly

There are quite a few YouTube videos. Really quite a lot, actually. So trying to *find* one is like looking for a movie in a haystack.

Here's a tip that you'll find helpful: Remember the magic word *allintitle* (as in "all in title").

If you type that before the search words in the Search box, YouTube finds only videos with those words in the *titles*.

For example, searching for "funny cats" finds tens of trillions of cat videos, even if those words don't appear in their titles. (The titles might be "Hilarious Kittens," "Yet Another Cat on the Piano," "Mittens Learns Her Lesson," and so on.)

But if you search for "allintitle: funny cats," then the *only* videos YouTube finds for you have the words "funny" and "cats" in their *names*.

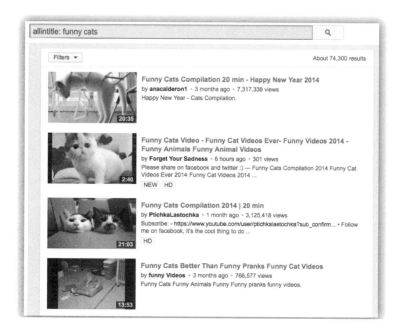

The secret game of Snake

When your YouTube video freezes because your Internet connection is gasping for breath, let's face it: That's a bummer. And it's boring.

Fortunately, the cheerful programmers at YouTube have given you something to *do* during that time: a game.

Next time you need the kind of excitement that only an arcade game can supply, *pause* the YouTube video.

Then press the ← and ↑ keys simultaneously.

Suddenly you're playing a game of Snake, superimposed on the paused video. (It doesn't *always* work; just accept that fact that Snakey's a little flaky.)

Funny Cats Compilation Part 1

The object of the game is to direct the snake around the screen with the arrow keys. But the longer you play, the longer the snake's body grows—and if you crash into the edge of the frame or cross your snake's body, the round is over.

Literally *minutes* of fun.

Never read a
movie critic again

For decades, we had only one way to research which movie to go see: checking the reviews in newspapers, magazines, TV, or radio.

The problem with that is that everyone is *different*—especially critics. You're putting stock into the opinion of *one single person,* who may have had a rotten day, who may be coming down with the flu, who may have gotten into bar fights with the movie's director back in college. You just don't know.

What you *should* do when choosing a movie is research it at RottenTomatoes.com. This site shows you the *collected* opinions—not only of all the critics but of all the normal people who've already seen the movie.

That's incredibly powerful. Yes, different people have different opinions. Yes, there are weirdos out there. But this site averages them all together, so you get a *really* good picture of the movie's quality.

Each average is expressed as a percentage. If you spot a movie with, say, 98 percent popular consensus, you're almost certain to love it. If you see one under 50 percent, you should probably save your money.

Rotten Tomatoes shows you very clearly that a critic's review and the public's review of some movies are *wildly* different, as you can see on the next page.

Another great source of guidance is IMDB.com (the Internet Movie Database). There you can see the averaged ratings of hundreds of thousands of people—and read their individual reviews. It's fascinating stuff.

The Boondock Saints (2000)

TOMATOMETER | All Critics | Top Critics | AUDIENCE

20% A juvenile, ugly movie that represents the worst tendencies of directors channeling Tarantino.

Average Rating: 4.2/10
Reviews Counted: 25
Fresh: 5 | Rotten: 20

92% liked it

Average Rating: 4.2/5
User Ratings: 332,317

TRAILER

PLAY HD TRAILER

How to cancel your cable TV and use the Internet

E very year, another couple of million Americans become "cord cutters." They cancel their cable TV or satellite service. Instead, they rely on TV shows and movies delivered from the Internet, for a fraction of the price.

Netflix and Hulu, for example, each offer an $8-a-month service that lets you watch all the TV and movies you want. Unlimited. On demand. Cancel your $100-a-month cable service, replace it with an $8-a-month service—you can see the appeal.

Now, video services like these don't have everything that's on TV. You lose sports, news, game shows, and sitcoms.

But you gain instant access to entire seasons of beloved TV series like *Modern Family, Breaking Bad, Arrested Development, Desperate Housewives,* and so on. Plus, on Netflix, you'll get instant, unlimited access to 20,000 movies—usually a year old or more, but brother, what a deal! All you want for $8 a month.

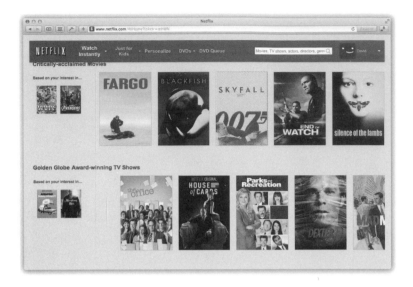

You can watch the shows on your computer, tablet, and phone. Most new TVs have Netflix and Hulu Plus built in, too (you still have to pay the $8 a month), so you can watch on the big screen—or you can buy a $35 gadget called a Google Chromecast. It adds those services to *any* HDTV.

The question is, what do you use TV *for*? If you use it just to entertain yourself in the evenings, then you're probably a good candidate for "cord cutting."

If nothing else, call your cable company and *say* you want to cancel. You'll be transferred to a special Department of Persuading People Not to Cancel Their Service—and they'll offer to lower your cable bill a *lot* if you agree not to leave.

Chapter 14:
12 Free Services You'll Adore

The Internet is big, dude. Bigger than big. There are *billions* of Web pages—so many that even Google can't search more than about 5 percent of them!

Let's admit it: A huge, seething majority of it is garbage. But logic tells us that in any ocean of material that vast, you'll find some real treasures. The purpose of this chapter is to make sure you're aware of them.

Store, display, and find photos online

I f there were a fire, what are the first computer files you'd grab? A lot of people would say, "My photos." So then look this page straight in the eye and answer the question: At this very moment, do you have an up-to-date backup (safety) copy of your digital photos?

Didn't think so.

That's why you should sign up for a free Flickr.com account. It gives you *one terabyte* of storage for backing up, displaying your pictures. How much is a terabyte? That's a thousand gigabytes. That's enough room for 650,000 full-resolution pictures. *That* oughta be enough to get you through your kids' first couple of birthday parties.

You can group and organize your photos in all kinds of different ways. You can also specify who's allowed to *see* your different albums: your friends, your family, everyone, or nobody but you.

The hassle is uploading all your pictures to Flickr; if you have a lot of them, it takes a long time. This Web page lists all the different ways to do that: *https://www.flickr.com/tools.*

You can send pictures directly to your Flickr account from within popular photo-organizing programs like iPhoto, Picasa, Aperture, and Adobe Lightroom. You can also send pictures from your phone or tablet. In fact, you can even *e-mail* pictures to your Flickr account, which is cool.

But you can also just drag folders of pictures right off your Mac or Windows desktop into the Flickr window, like this:

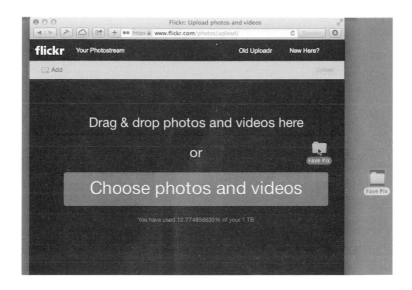

Never visit a bad restaurant —or bad hotel—again

Never again will you choose a restaurant based on how it looks from the street. And you've stayed at a gross, creepy hotel for the last time, too.

The beauty of the Internet is how it harnesses the wisdom of the masses. It's perfectly possible to look up a restaurant or a hotel and read *hundreds* of reviews from people who've eaten or stayed there. If there are cockroaches, you'll know it immediately.

The sources of all this wisdom: Yelp.com (for restaurants and stores) and TripAdvisor.com (for hotels and attractions). It also has apps for your phone that make it easier to look things up.

There's no excuse to buy a meal or a hotel room, blind and uninformed, ever again.

Make reservations from your phone—and eat free

OpenTable is a fantastic Web site (and, of course, phone app) that lets you make restaurant reservations. No waiting on hold, no trying to make yourself heard on the phone, no having to ask for their available times (and then finding out they have none).

You just tap in what kind of food mood you're in. You instantly see which reservation times are available.

Better yet, every time you use OpenTable, you accrue frequent-eater points—and you can cash them in for free meals at the same restaurants you've been enjoying.

The only downside: OpenTable is pretty much a big-city thing, and not all restaurants participate (they have to pay for the privilege). But for the places that do, there's nothing like it.

Tap into travel agencies' flight databases

Flight information is no longer the private domain of airlines and travel agencies. You can look it up, too.

Plenty of Web sites let you search for flights and compare their prices (Expedia, Travelocity, Orbitz, and so on). But Hipmunk is by far the easiest to figure out. You specify when and where you're going …

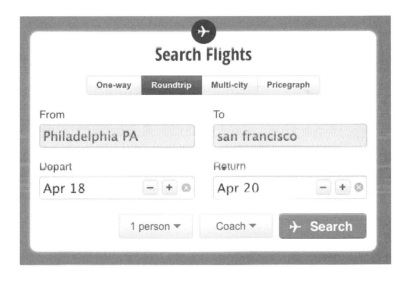

… and then view the results as a clear visual timeline of flights. You can sort them by price, duration, departure time, arrival time, or agony (high price, long flying time, and a lot of stops).

$892 roundtrip	US Airways		🛜 US Airways					
$842 roundtrip	United			United				
$892 roundtrip	US Airways			🛜 US Airways				
$812 roundtrip	Virgin America					🛜 Virgin America		
$812 roundtrip	US Airways					🛜 US Airways		
$812 roundtrip	United					United		
$892 roundtrip	US Airways						🛜 US Airways	

You can buy the tickets on the spot if you so desire. It's pretty fantastic.

Find out a flight's terminal, gate, and actual arrival time

If it's your job to pick someone up at the airport—or if you're flying—you should know about FlightAware.com. It's an incredible source of information about any flight in the air.

Examine the information about the flight, and you'll see such amazing information as:

- Where it is over the country right now, on a map

- What time it *really* lands (compared with its recent history)

- What terminal and gate it's landing at (and which it took off from)

- Exactly how long it's been in the air—and how long it has left to fly

- Its altitude

- What kind of plane it is, and even:

- The average ticket price the passengers paid!

This information is drawn directly from live FAA data, so it's far more current and truthful than, for example, what the airline will tell you. Now you know the secret.

Free classifieds
that work scarily well

Craigslist.org, it has been said, is killing the American newspaper. It has taken over the function of what used to be called classified ads.

Craigslist lets you put up any kind of ad free and instantaneously. For sale, help wanted, employment sought—everything—in every city of many countries.

And you would be amazed at how many times people near you are willing to pay money for your old junk, how many great candidates you get for a job you're trying to fill. A *lot* of people look at Craigslist every day.

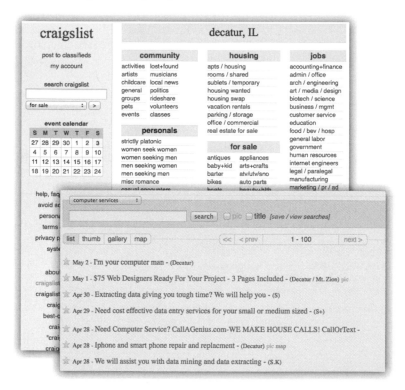

And since it's free, and easy, keep it in mind the next time you want to buy, sell, rent, hire, work, or date.

Find the cheapest gas near you

The Internet knoweth all things, because the Internet collecteth information from millions of people in real time. Including, surprisingly, the prices of gas at all of the gas stations near you, as reported by fellow drivers.

At GasBuddy.com, specify where you are, and you'll instantly see a list of the lowest gas prices nearby.

Search Gas Prices Show Search Options

🔍 Chagrin Falls, OH   Search

Lowest Regular Gas Prices in the Last 24 hours

Regular Gas	Midgrade	Premium	Diesel Fuel
Price	Station	Area	Thanks
3.69 update	Shell 8501 East Washington St near OH-306	map Chagrin Falls	BkGmo 🚗 4 minutes ago
3.75 update	Marathon 4975 SOM Center Rd & Miles Rd	map Chagrin Falls	kwwcjkm 🚗 3 hours ago
3.79 update	Marathon 130 Bell St & Philomethian St	map Chagrin Falls	KSKHarrold 3 hours ago
3.79 update	BP 20 E Washington St & S Main St	map Chagrin Falls	kwwcjkm 🚗 3 hours ago

This idea, of course, screams out for a phone app—because the time you're most likely to need it is when you're driving around. And, of course, there *is* a phone app. Also free.

Send enormous files

E-mail isn't a good way to send big files. Most e-mail systems don't accept file attachments bigger than about 10 megabytes.

Dropbox is a good option, but the free account may not leave you enough space—and you may not feel like installing *software* just to send someone a big file.

What you want to use in this case is SendBigFiles.com. (SendThisFile.com works similarly.) It could not be simpler: Choose the file you want to send, specify the e-mail address of the lucky recipient, and send.

The service is free, although you get faster transfers, and larger maximum file size, if you're willing to pay. Most of the time, though, you'll be grateful just to get the job done. (If you have the latest Mac operating system, Yosemite, you don't have to worry about any of this; you can send giant attachments in the standard Mail program without worry. Your recipient just clicks a link in the e-mail to download your file.)

Free conference calling

FreeConferenceCall.com is pretty amazing. It's exactly what it sounds like: free conference calls. It's ideal for setting up business calls, conducting group interviews, planning family reunions, and so on. Each call can handle up to 1,000 people for up to six hours.

You go to FreeConferenceCall.com. You sign up with your e-mail address. It gives you a number to dial and an access code. Everybody in the group calls in, enters the code, and boom: free conference call.

The pleasures of a good GoogleFight

Here's a great way of settling the next family argument: have a Google Fight.

At GoogleFight.com, you enter two search terms that you want compared. Abe Lincoln vs. George Washington? Bush vs. Gore? Spinach vs. kale? Shakespeare vs. Michael Crichton?

Click the Fight button. After a quick animated battle, you get to see which one is more popular/significant as determined by the number of mentions on the Internet. (Why are the labels in French? Never mind.)

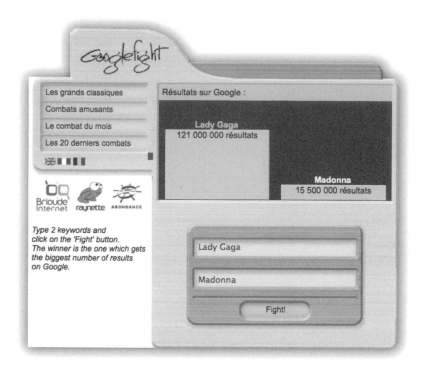

Type 2 keywords and click on the 'Fight' button. The winner is the one which gets the biggest number of results on Google.

Play any song, band, or album you want, free

I t's called Spotify. All the kids are doing it.

Once you sign up for a free account, you can type in the name of any song, band, or album, and listen to it—instantly. It's the world's biggest music store, and you own it. Go nuts.

Settle an argument. Play a new favorite song for your friends. Teach your kids about music you used to love.

You can also build playlists of similarly themed music—great for a party, wedding reception, date night, and so on.

Spotify requires you to install software onto your computer or phone, and it treats you to the occasional ad (unless you sign up for one of the monthly-fee plans).

If the music you want to hear is more recent, here's another, unexpected idea: Look it up on YouTube. All recent popular music is there, free to play, usually in the form of music videos. Great classical music pieces are there, too, and even songs from movies and Broadway shows.

Rewind time to see an older version of a Web site

Web pages aren't like magazine pages; Web pages change all the time. And once somebody updates a Web page, most people probably assume that the older version is lost forever.

That's not true. An astonishing Web site called the Internet Archive Wayback Machine actually *stores* old versions of most Web pages. You can choose on a calendar which old version you want to call back from the dead.

To use it, go to *archive.org/web* (no "http:" or "www" on that address). In the box, type in the address of the page you want to see (for example, *www.nytimes.com*); click Browse History.

If the Wayback Machine does, in fact, list that site, you now see a calendar, going back years, showing the dates for which it has a captured snapshot. (Some sites, including Facebook, have asked to be excluded from the Wayback Machine. Older versions of those sites really *are* lost forever, at least to you.)

Click the date you want to see, and marvel at how times, design, and computer screen sizes have changed.

Chapter 15:
10 Fantastic Phone Apps to Install Right Now

The best part of a smartphone (an iPhone or Android phone) is that it lets you install *apps* (small software programs). With the right app, your touchscreen phone can turn into all kinds of *other* gadgets: camera, scanner, musical instrument, heart-rate monitor, and so on.

But if you've a recent arrival in the land of iPhones or Android phones, you might have no idea where to start. There are, after all, over a million different apps to choose from; trying out all of them would take, you know, a whole weekend at least.

So here are ten of the greats (each for iPhone or Android). Ten apps that are so useful and so well done, they belong on almost everyone's phone.

Google Maps

Most people probably think of Google Maps as a GPS navigation app. And yes, it can indeed turn your phone into something like the GPS receivers people stick to their windshields with suction cups. You tell it where you want to go, and a synthesized voice gives you driving directions.

But Google Maps is almost infinitely better than one of those windshield gadgets. First, because it's almost supernatural at its ability to figure out what destination you want. Looking for Whole Foods? Once you've typed *wh,* the app proposes Whole Foods Market. Want guidance to the Empire State Building? *Emp* is about all you need to type.

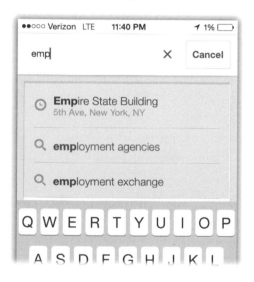

Another big advantage: Google Maps is constantly being updated to keep up with the changing roads of the world. It's *really* good.

Only some of this app's usefulness is related to driving. All of its stunts, however, begin with specifying a place—a street address, the name of someone in your address book, a business name, even an intersection. To begin, tap in the Search box at the top (A in the diagram on the next page); start typing your destination's name as described above. (You can be vague; *125 St, 125th street,* and *125 street* all work.)

Google shows a list of search results; tap the name of the one you want. Now the fun begins:

- **Get driving directions.** Tap the blue car icon (B). Now you arrive at the "How do you want to get there?" screen. Tap your mode of transport (driving, public transportation, biking, or walking) at the top (C). Then scroll up to see the different proposed routes. Almost always, the first one is the quickest. Tap "Start navigation" to begin the turn-by-turn spoken guidance.

Just in case you were reading quickly, yes: this app can actually give you *detailed bus or subway* instructions for getting someplace, including how much time each leg takes.

And by the way: It also color-codes the roads to show you how bad the traffic is on your route. And it offers to guide you around accidents and construction sites.

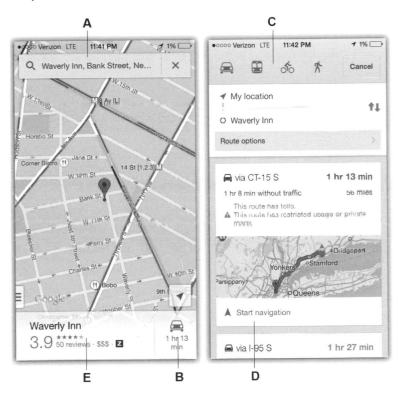

- **Call a place.** Google knows about millions of stores, restaurants, and so on; it's like a global Yellow Pages. To call a place, specify its name as described above, then tap its information banner at bottom (E). There's its address, Web site, photos, and so on—and, best of all, a Call button. Tap it to call the place. You'll never even know or care *what* the phone number is; you'll just be connected, in a hurry. Infinitely faster than looking up the number online.

- **Investigate restaurants.** In the Search box, specify what you're looking for: Italian restaurants, sushi, whatever. Google Maps shows you, at bottom, the closest match, complete with its customer rating and price range (E). Drag the banner sideways to see the capsule views of other restaurants in this category.

When you see an eatery that looks good, tap the banner to call, or get driving directions, or read customer reviews or Zagat Guide reviews, or even see photos of what the place looks like inside. (Rumor has it that in the next version, the app will actually let you taste samples of each restaurant's dishes, served on little pop-out toothpicks.)

Of course, this trick works with all kinds of businesses, but restaurants are the most commonly sought examples.

HealthTap

Wouldn't it be cool if you could ask a doctor something—right now, from home—and get an answer for free? That's what HealthTap (free) lets you do. Type in any non-emergency medical question, and you'll get an answer from a doctor (or several, which, if they back each other up, boosts your confidence), usually within a day. It's rather fantastic.

Your question, and the answers, are visible to everyone, although you can sign up anonymously if you like. If you prefer to ask a question privately, and to a *particular* doctor, you can pay 10 bucks.

FlightTrack

I f you fly, you need this app ($5). Once you tell it what your
flights are, FlightTrack shows you every detail about them:
gates, time delayed, airline phone numbers, where the flights
are on the map, and so on. You'll quickly discover that it knows
more than the airline gate agents do—and knows about changes
much sooner.

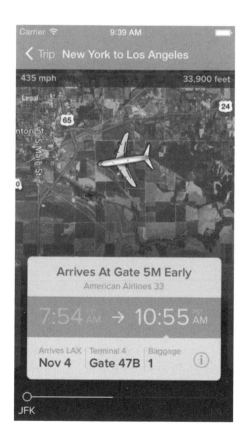

SoundHound

Sometimes, an app is so good, it approaches magic—and this is one of them. It identifies songs for you by *listening*.

For example, suppose some song is running through your head, but you can't remember what the heck it's called. In this app, you can tap the Listen button and *hum it*. And presto: the app tells you the name of the song and performer, displays the lyrics—and, of course, offers a link that lets you *buy* the song. (Its rival, Shazam, is better known—but it doesn't do that humming-recognition thing.)

Or if you're in a restaurant hearing a song you like, or whose singer you can't identify, you can tap Listen and then hold the phone up in the air. The app sniffs the sound and then tells you what song is playing, even if the restaurant is noisy. It's a great way to satisfy your curiosity, discover new music, or settle important family arguments.

Flipboard

This free app offers you a huge table of contents of news sources from the Web: sports, politics, entertainment, music, movies, technology, and so on. Once you've indicated your preferences, presto: You've got a gorgeously illustrated digital magazine right on your phone, crafted to your own peculiar tastes.

Flipping through your custom publication is perfect for those dead spots in your day—when you're waiting to see the doctor, waiting for the plane to take off, or waiting in line at the Department of Motor Vehicles.

Flixter

The ultimate movie showtimes app. Right on the home screen, you see a list of movies playing right now, complete with the ratings (as in, PG or R), ratings (as in, "4.5 stars"), running length, and who's starting. If one looks interesting, you can tap it to watch its trailer (preview) and read a description. If it *still* looks interesting, another tap shows you which theaters near you are showing it, and what are today's showtimes.

You can also flip it inside you. That is, you can start with a list of theaters near you, and see what's playing at each one. Mr. Moviefone is dead. Long live Flixter!

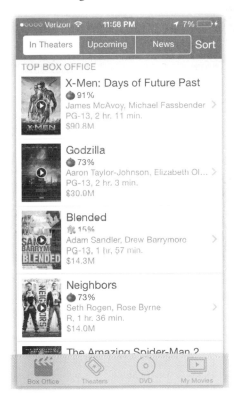

JotNot Pro

I t's a scanner. Really.
Point your phone at any document—a receipt, an article, a legal brief, a page from a book, a piece of sheet music, blueprints for a Russian spy satellite—and this $3 app scans it, just as an actual scanner would.

It's a lot like taking a picture of something, except that the app offers tools for straightening the "scan," fixing distortion, adjusting the contrast, and so on. And it keeps multiple-page documents together.

When it's all over, you can send the finished scan by e-mail as a PDF document, print it wirelessly, drop it into your Dropbox, or even (for a fee) send it to someone's fax machine. (You know the one sitting next to a kerosene lamp and butter churn.)

Uber

Uber, operating in over 100 cities, is a simple, brilliant idea that changes everything—making some people delighted and others livid.

It calls cars to take you places. You can choose an Uber (which summons a professional driver working for a car service) or, for less money, an UberX (a *non*-professional—just a normal person in a normal car who's got some time to drive people places in hopes of making some extra cash).

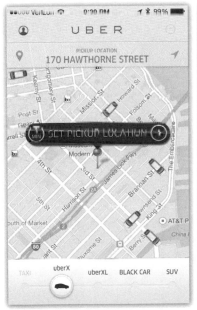
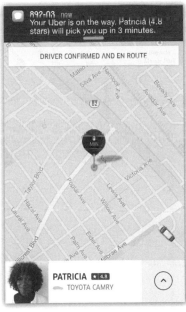

You should think of Uber any time you'd ordinarily think of getting a taxi or even a rental car. It's just incredibly simple. When you open the app, you see a map that shows where the closest cars are. One tap, and the nearest one is on its way to pick you up. You see the driver's photo, phone number, and rating from previous passengers.

When you arrive, you don't have to fuss with cards, cash, or tipping. You just say "bye!" and get out. The billing is handled automatically.

It's a life changer.

Pandora or Spotify

Pandora is like a radio station that actually cares about you. You choose a "seed" song or band that you like. This free app then uses the details of that song (the beat, the tempo, the instrumentation, the kind of singer, and so on) to play other songs for you that more or less match the first one. As each song plays, you tap a Thumbs Up or Thumbs Down button—and Pandora learns more and more about what kind of music you like. Pretty soon, it plays *only* songs that you really enjoy.

You hear an occasional radio-style ad. Paying $5 a month gets rid of the ads.

Then there's Spotify. This Internet music service is described on page 308, but there are some interesting differences in the phone app.

If Pandora is like a you-specific radio station, playing endless music in styles you like, Spotify is like a free music *store*. You have more control. You can play almost any album in the world—for free, on demand. Or listen to any particular performer. Or listen to ready-made playlists of music in every conceivable category (Romantic Guitar, Pulsing Workout Music, Ethiopian Harmonica, and so on).

As with Pandora, the free service has some limitations: you hear occasional ads, you can't call up a particular *song* by name, and you can't control the playback *order* of songs within an album or performer. Paying $10 a month eliminates the ads and lets you listen to any individual *song* (not just albums or performers).

Dictionary.com

It's a little weird that your do-everything phone doesn't have something as important as a dictionary, don't you think?

This free app is just what you expect: a dictionary. It contains 2 million words, with definitions.

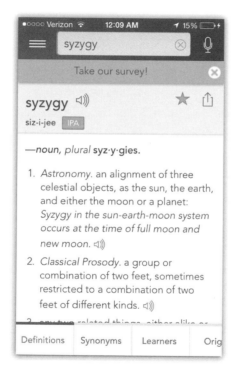

The Usual Suspects

The suggestions on the previous pages are ingeniously designed to introduce you to terrific apps you might not have heard of.

But you should also install the free apps for any online services you use on your computer: Facebook, Twitter, LinkedIn, Flickr, Evernote, Yelp, Dropbox, and so on. These free services become even more useful when you have them right on your phone. For example, you can check Facebook or Twitter whenever you have a few minutes to kill between obligations, or you can use Flickr to call up any photo you've ever taken. It's as though your phone has become the world's longest fold-out wallet photo holder.

Part 4

Social Networks

Chapter 16: **Facebook**

Facebook is the second most popular Web site on the planet. About 1.3 billion members—about a sixth of the earth's population—visit it every month.

The idea is simple: You tell Facebook who your friends, family, and colleagues are. If someone agrees that she knows you, you two have just "friended" each other.

Your Facebook page is like a scrolling newspaper of news and photos posted by these friends.

You may mutter that you're not getting involved with some cockamamy site whose purpose seems to be invading your privacy. Lots of people mutter that.

But in truth, you have complete control over your privacy on Facebook—and by avoiding it, you miss out on a truly wonderful way of staying in the lives of your friends and family.

Who sees what you post on Facebook

When you open Facebook.com, the box at the top invites you to type something. Say what you're doing, what you're feeling, what interesting thing you found online. Announce a birth, a marathon, a new job.

Before you click the Post button, though, don't miss the pop-up menu next to it. There you specify *who can see* what you've just typed: Public, Friends, Only Me, or Custom (a group of people you specify).

Be warned, however: The choice you make right now will *stay* made for future posts. If you change it to say Public now, it will still be set to Public the next time you type something on Facebook—unless you change it again.

The best recommendations on the Web

Google is great for finding *information* about things. But when you want a *suggestion,* use Facebook.

Ask what you should do with your parents (or your kids) when they come to visit next weekend. Ask what's a good model for a kid's first cell phone. Ask what's a good drum set for a ten-year-old. Ask where to go for sushi in Albuquerque.

David Pogue
January 17, 2013

OK. Three-day weekend coming up. I've got the boys (ages 15 and 8). Who can suggest a great weekend getaway, driveable from Connecticut? I could use some leads! Thank you!

Like · Comment · Promote · Share   👍 36   💬 146

Roz Mandelcorn THE INAUGURATION in D.C!!!
January 17, 2013 at 6:24pm · Like · 👍 3

Heather Robinson Jay Peak has skiing plus a water park!
January 17, 2013 at 6:27pm · Like

Paul W. Lally Bring them to Newport, RI. Huge mansions, great restaurants, beautiful harbors and beaches, centuries of history.
January 17, 2013 at 6:30pm · Like · 👍 3

Alan Foster Mark Twain's house in Hartford.
January 17, 2013 at 6:43pm · Like · 👍 1

Jim Kanter Snowboarding in the Poconos.
January 17, 2013 at 6:51pm · Like

John Cooksey That the Acela express train to DC and see the Smithsonian Museaums
January 17, 2013 at 7:44pm · Like · 👍 2

Marlene Greenman Heller Gettysburg PA. Take a self-guided tour. Go to Millers for a meal. Shop at the outlets. Visit a pretzel factory, see the Amish.
January 17, 2013 at 7:45pm · Like · 👍 2

The genius part is that the only answers you get come from *people who know you*. Your friends, family, colleagues. People who, presumably, have similar taste and experiences—or at least know about *you* and can guide their suggestions accordingly. You'll be pleasantly surprised at how well this system works.

The Facebook messages you've been missing

You might be in for a shock. You're about to discover a whole lot of e-mail that you've been missing. Depending on how long you've been on Facebook, you might have missed important messages going back years, with long-since-expired offers,

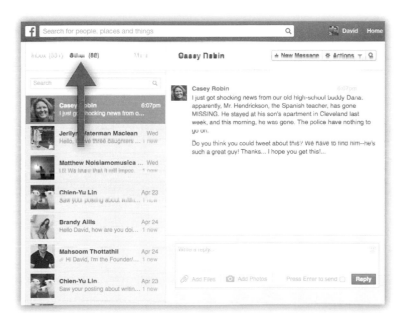

invitations, condolences, and congratulations. Ex–romantic interests, old bosses, school friends.

To see this folder for yourself, click the Messages button on the left side of your Facebook page. There, next to the boldfaced word Inbox, is the light-gray word Other. Click it to see your hidden stash of messages.

Disaster averted.

Why would Facebook create a folder full of mail that we never see?

Facebook has an interesting message policy: If you try to contact someone who's not one of your Facebook friends, a box appears that offers you a choice. You can pay $1, or you can send your message to the recipient's Other folder. (This setup is intended to thwart spam; it ensures that the only messages in your Inbox have come from your friends.)

You don't get any notification when new messages arrive in your Other folder, though; it's up to you to remember to check it.

Facebook's secret keyboard shortcuts

Without the transcendental ability to peer into Mark Zuckerberg's mind, you'd never guess that touching certain letter keys performs all kinds of useful stunts in Facebook.

Type each letter key by itself—no need to add Ctrl, or ⌘, or whatever.

- J Tap this key to scroll neatly from headline to headline in your "news feed." (Press the letter K key to jump back up again.)

- **L** Clicks the Like button for whatever you're reading.

- **C** Jumps into the Comment box, so you can type a remark about the current story.

- **S** Means, "Share this story" (re-Post it on your own timeline).

- **P** Drops your cursor into the status box so that you can type, and Post, a new Facebook status message.

- **/** Press the slash key to open the Search box.

- **?** Opens a little cheat sheet that lists all of these keyboard shortcuts!

You can also use keyboard shortcuts to jump among Facebook's various screens. The next list shows the keyboard shortcuts for most Windows browsers (like Internet Explorer and Chrome). If you're using a Mac, press the Control and Option keys instead (for example, Control Option-1).

And if you're using Firefox for Windows, press the Shift *and* Alt keys with each number

- **Alt+0:** Help screen.

- **Alt+1:** Your Facebook home screen.

- **Alt+2:** Your profile page.

- **Alt+3:** Your list of friends.

- **Alt+4:** Your Facebook message inbox.

- **Alt+5:** Notifications.

- **Alt+6:** Facebook settings.

- **Alt+7:** Your Activity Log.

- **Alt+8:** Facebook's Facebook page.

- **Alt+9:** Facebook's Terms and Policies.

- **Alt+m:** Start composing a new Facebook message.

Chapter 17: Twitter

There are actually *two* 800-pound gorillas of the social media world. Facebook is one; the other is Twitter.

Who would have thought that a service whose articles are limited to *140 characters* would become a mega-success with hundreds of millions of members?

But that very limitation is precisely what inspires Twitterites. The length limit forces you to be concise, encourages you to be witty, permits you to skim through dozens or hundreds of messages ("tweets") whenever you get a moment.

Twitter is like the transcript of a global cocktail party. Headlines, gossip, jokes, and opinion tear through Twitter like shock waves. News breaks there hours before the "real" news organizations even hear about it.

Twitter is not easy to dive into. But you'll find it easier than most people do—because these are the Twitter Basics.

Whom to follow

If you sign up for your free Twitter account and then just stare at the screen, you'll be bored out of your mind. You'll receive only the tweets of people you've *followed*—people whose tweets you've asked to get.

That's why, when you first sign up, Twitter offers you a few lists of people it thinks you might want to follow: famous people, people in fields like photography and music, and people you *know*, based on your online address books (Gmail, Facebook, and so on). By the time you're done, you'll already be following at least 15 people or so.

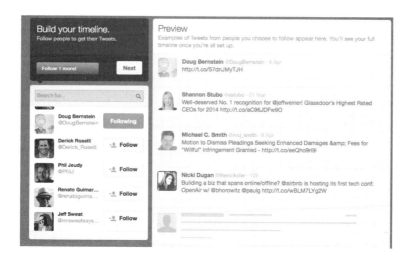

You're ready to start doing Twitter. Log in to Twitter.com whenever you have a moment, and just scroll through the accumulated messages. Over time, you'll build a bigger list of interesting people to follow; you'll *un*follow boring or offensive people; and before long, you'll find yourself right at home.

What all the symbols mean

Some tweets are self-explanatory, interesting, or amusing, and require no decoding. Here's one like that: "Eating outside on a blanket is no picnic, let me tell you. Oh, wait. It actually is."

Others, however, can be baffling at first, thanks to the shorthand people have developed to save space. (It's that 140-character thing again.)

Suppose, for example, that you see a tweet like this:

"It is hard to spell or type properly. RT@snosk What's Wrong with Sentimentality? http://j.mp/1n1UTwj"

In that tweet, you can find *three* Twitter conventions that nobody tells you about:

- **@snosk.** Everybody on Twitter has a name. Sometimes it's obvious who it is —@taylorswift13, for example, is singer Taylor Swift. Other times, it's a little joke; defensive NFL lineman Warren Sapp goes by @QBKilla ("quarterback killer").

  So what about that @ symbol? That just designates a Twitter handle. Any word preceded by @ is a Twitter name.

- **RT @snosk.** RT stands for *retweet*. In other words, this guy is *repeating* (or, all right, *retweeting*) something somebody *else* said—in this case, something @snosk said.

  It's polite to use RT when you're repeating someone else's tweet, so the masses don't think your trying to pass off *his* brilliance as your own.

So in the tweet above, **@snosk asked, "What's Wrong with Sentimentality?"** And this tweet answers the question humorously: "It is hard to spell or type properly."

- **http://j.mp/1n1Utwj.** That's a link to a Web page. That's very common in tweets. People use Twitter to share great stuff they've found online. (In this case, clicking on that link would have taken you to a magazine article called, yes, "What's Wrong with Sentimentality?"—the item that started this whole thing.)

- -

A crash course in hashtags

Many tweets end with creative or peculiar notations like this: #oneliners, or #nyyankees, or #gotwhatideserved. It's a little baffling, to be sure.

These are called *hashtags*. That's Ancient Geek for "labels that people can use to find tweets on a certain topic."

Suppose, for example, you want to know what everyone in the whole world is saying about the World Series. You can go to Twitter.com, click in the Search box, and type #worldseries. When you press Enter, you'll see *nothing* but tweets that mention the World Series. Or, rather, you'll see tweets that contain the hashtag #worldseries. People tag their tweets with those words *just* so people can find them later.

Another case: You might be reading along through your tweets—and find an intriguing hashtag (say, #lousypun). You can click directly *on* that hashtag to see tweets that have the same hashtag.

Bernie Fratto @BernieFratto · 44m
It is the 30th anniversary of the '84 #WorldSeries & the #Padres are exacting revenge on the #Tigers -Yeah, that might be a lame #tweet #MLB
Expand   ↩ Reply ⭣⭡ Retweet ★ Favorite ··· More

Rick Stevens @Rickstevens24 · 46m
@Mariners loving this team!! #WhynotMariners!! #WorldSeries
Expand   ↩ Reply ⭣⭡ Retweet ★ Favorite ··· More

Great Davey's Ghost @Ace_Gagnon · 53m
Took the #Rangers 12 innings to beat the #Astros 1-0 and they're acting like they won the #WorldSeries. #MLB
Expand   ↩ Reply ⭣⭡ Retweet ★ Favorite ··· More

James Robert Damude @Jamestrailmix · 1h
The worst part of the fixed 1919 #WorldSeries wasn't cheating or gambling but the suspicion they put into young kids hearts. #baseball
Expand   ↩ Reply ⭣⭡ Retweet ★ Favorite ··· More

James Robert Damude @Jamestrailmix · 1h
The 8 players who help fix the 1919 #WorldSeries were found innocent when transcripts of confessions made to Grand Jury vanished #baseball
Expand   ↩ Reply ⭣⭡ Retweet ★ Favorite ··· More

Those, at least, were the original purposes of hashtags.

These days, though, people make up hashtags that nobody will realistically search. Instead, they use hashtags sardonically, as little phrases that *comment on* the tweet.

A typical example: "The TV screens on my flight to Hawaii stopped working TWICE! #firstworldproblem." Meaning, "OK, I realize that sounds like the petty frustration of a rich guy."

Or you might read: "I could definitely afford a new car— if I win the lottery. #notholdingmybreath." Get it?

And now, people even use hashtags *outside* of Twitter. They sometimes even *speak* them, as in "hashtag Awkward!" or "hashtag FAIL!"

Here's where things get complicated

I f your tweet *begins* with somebody's Twitter handle, it will appear in that person's stream of incoming Twitter tweets, as you'd expect. But it will also appear in the Twitter streams of anyone who happens to be following *both of you*.

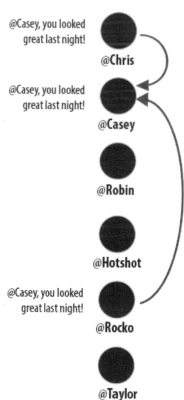

This diagram may make that clear. Then again, it may not. It's showing that if you address a message to @Casey, then @Casey will see it. But so will @Chris, @Rocko, and anyone else who's following both of you.

(Nobody else on Twitter will see any of this. But anybody *can* find it by searching for it.)

The moral of the story: If you want to say something privately to one person, send it as a direct message, described next.

Send a message privately

Ordinarily, anything you type into Twitter pops up immediately on the screens of anyone who's following you.

Often, though, you have something to say to only *one* person. A question, an update, a note of congratulations. Something of no interest to the masses.

In that case, put the letter D in front of what you have to say (either *d* or *D*), and then the person's Twitter name. Like this:

d @BarackObama Hello, sir! I really enjoyed your appearance on Conan.

That's called a *direct message,* and it goes directly to the person you've identified—on one condition:

That person must be one of your followers.

In other words, you can't send direct messages to people who *aren't* following you. They just won't get your note. Twitter invented this rule to prevent you from drowning in spammy tweets that clog your timeline. It is, however, a confusing rule. The Twitter global transcript is filled with people saying "@ FoxyChique: I can't DM you because you're not following me."

Twitter on your phone and tablet

Most people read Twitter on its Web site, Twitter.com. But one of the smartest things Twitter Inc. ever did was to let *other* people write programs that can send and receive tweets. There are Twitter apps for Mac, Windows, and every smartphone and tablet. (Actually, you can get tweets even on a *regular* cell phone, too—as text messages. That's why Twitter has the 140-character limit.)

These other apps have many more features. For example, they often let you see your Twitter world in *columns*. The first one shows incoming tweets from everyone you follow. The second one shows tweets that mention, or are intended, just for you. And the third shows private messages sent to you.

You are by no means a loser if you do all your tweeting on the Web site. This is simply a public-service announcement to make sure you know that Twitter is waiting for you in other places, on other gadgets.

Acknowledgments

And with that, you've reached the end of the book—the last of the 225 most important tech techniques. Here's hoping that at least a few of them will stick with you. May they send you out into the night with a renewed sense of confidence—or at least fewer wasted taps, clicks, and minutes.

In reverse chronological order, these are the delightful people who made this book possible:

At Flatiron Books: Jasmine Faustino, who made the experience wonderful; managing editor Eric C. Meyer, who humored my tweakiness on layout concepts; and publisher Bob Miller, who had the spectacular good taste to decide to publish this book on Flatiron's very first list!

At the Levine Greenberg Rostan Literary Agency, my friend, and the world's best book agent, Jim Levine.

At TED, Chris Anderson and Bruno Giussani, who invited me to speak at the 2013 conference. My topic may sound familiar; it was "10 Tech Basics You Think Everybody Knows (But They Don't)."

During this book's creation, I enjoyed the support and infinite patience of Jan Carpenter, Cindy Love, my team at Yahoo Tech, and my brilliant Brady Bunch of a brood: Kell, Tia, Jeff, Max, and Farley.

Above all, I owe a debt to my beautiful bride Nicki. Her love and encouragement carried this project all the way from, "You know what I should write someday?"—to the finished book in your hands.

Index

abbreviation-expander feature,
  Mac, 121–22
accents, typing of, 122–24,
  180–81
Adblock, 260–61
Adblock Plus, 260–61
AirDrop feature, Mac, 135–36
Airplane Mode, smartphone
  battery life and, 14, 16
  international travel and, 29
Amazon.com, 217
  books viewable on, 67
  "open box" specials by, 73
  site navigation on, 253–54
Android phones, 6–7. *See also*
  cameras, smartphone;
  phone app(s); smartphones
  app deletion on, 26–27
  apps for, 30–31, 34, 35, 43,
    66–67, 311–25, 342
  autocorrect tips for, 23–25
  automatic period trick on, 13
  battery life tips for, 15–16
  camera tips for, 31–32, 62–64
  directory assistance tips for, 26
  e-book apps for, 66–67
  international travel tips for,
    29–30
  keyboard tips for, 21–22, 43
  lost, 35–37
  password protection on, 37
  QR information codes and, 34
  recharging tips for, 13–14, 43
  redialing tip for, 17–18

ring muting on, 17
screenshots on, 28
setting adjustments on, 12–18
software versions of, 9
voice mail tips for, 20
wet phone salvaging and,
  18–19
Wi-Fi settings on, 32–33
Android tablets
  charging tips for, 13–14, 43
  keyboard tips for, 21–23, 43
  as picture frame, 47–48
  rotation lock on, 46–47
antivirus software, 147–49,
  215–16
Apache OpenOffice program, 207
Aperture program, 298
Apple. *See also* iPads; iPhones;
  iPods; Mac computers
  new product introductions by,
    4–5
  One to One lessons offered by,
    145
  proprietary advantage of, 111
Apple Mail, 189
  e-mail attachments in, 199,
    233–35, 239–42, 306
applications (apps). *See also* phone
  app(s)
  definition of, 7
  deletion of, 26–27
  phone, 30–31, 34, 35, 43,
    66–67, 311–25, 342
asterisks, e-mail use of, 237

autocorrection, smartphone,
23–25
auto-formatting, computer,
184–88
auto-on/off feature, Mac, 142–43
auto-reopening (of documents),
Mac, 115–16
Avast Free program, 149
AVG Free program, 149

background updating,
smartphone, 16
Backspace key, functions of, 147,
156, 245–46
backups of computer, 221–23
online, 105–6, 219–20
banner dismissal, iPhones, 38–39
Barnes & Noble e-books, apps
for, 66–67
batteries, smartphone
extending life of, 15–16, 43
recharging, 13–14, 43, 44–45
BCC feature, e-mail, 235–36
Best Buy, recycling by, 6
bitly.com, 250
browsers, Web, 7, 243
address bar of, 244–45,
247–48
Back button navigation on,
245–46
privacy modes on, 263–64
search options on, 282–84
tab use in, 251–52
URL shortening and, 249–50
business directories, online,
278–79

calculators
on computers, 126–27

Google as, 276–77
cameras, digital
flash tips for, 51–54
focus/exposure lock on, 62–64
photo cropping and, 61–62
recovery of deleted photos on,
59–60
release dates for new, 4
self-timers on, 31–32, 58
sensors in, 54–56
sharpness of picture and,
54–56, 58
shutter options on, 50, 60–61
tripod tips for, 56–57
cameras, smartphone, 71
focus/exposure lock on, 62–64
self-timer apps for, 31–32
Card Data Recovery program, 60
CardRescue program, 60
Character Map program, 181
chargers, gadget
labels for, 73
Chrome, 7, 33, 156, 245–46, 251
privacy settings in, 263
Classic Shell app, 162
clicking. See also mouse devices
definition of, 7
right, 78–80, 155–56
computers. See also keyboards;
mouse devices; shortcuts,
keyboard; specific computers
auto-formatting options on,
184–86
backups of, 105–6, 219–23
data recovery on, 60, 86–87
declining use of, 12
file alphabetizing on, 99–100
file deleting on, 86–88, 134
file locating tips for, 89–90,
93–96
file saving on, 93–97

file selecting on, 89–92
file storage/transfer services for, 105–6, 135–36, 219–23, 235, 298, 306
forced closure of programs on, 109–10
laptop, 77, 112, 142–43, 218
menu tips for, 107
mouse functions on, 78–80, 83–84, 102–3, 134
music playing/pausing on, 103–4
navigation between open programs on, 101–2, 158, 168–70, 196–97
password tips for, 112, 149–52, 217–18
refurbished, 5–6
scrolling tips for, 97–98, 113–14, 254–55, 286–87
shortcuts, keyboard, for, 8, 108–9, 140–42, 160, 163, 179–80, 184, 203, 245–48, 287–88, 332–34
sleep mode on, 77, 142–43
software versions of, 9
tech support for, 104–5
text deleting on, 84–85, 210–11
text selecting on, 83–84, 92–93
Trash/Recycling Bins on, 86, 88, 177–78
"Undo" command on, 80, 211
"Yes" and "No" keys on, 81–82
conference calling, free, 307
conversions, online, 272, 281–82
Craigslist.org, 304–5
cropping, photo, 61–62
cursor, enlarging of, 164–65

Dashlane program, 218
data recovery
  on cameras, 59–60
  on computers, 60, 86–87
data, smartphone
  battery life and, 16
  "pushing" of, 16
date stamps, computer, 159–60
Del/Delete key, Windows, 147
deletion
  of files, 86–88, 134
  of text, 84–85, 210–11
dictation, text, 125–26, 165–68
dictionaries, online, 266, 278, 324
Dictionary.com app, 324
directory assistance, free, 26
discount coupons/codes, 6, 72. *See also* free services, Internet
drag, definition of, 7
Dragon NaturallySpeaking program, 165
drive.google.com, 208
DriveSavers program, 87, 209
Dropbox, 105–6, 235, 306
  app for, 325

earbuds, Apple
  control functions on, 32, 70–71
  photo tips for, 32, 71
e-books, apps for, 66–67
electrical outlets, tips for, 68–69
electronic signatures, 118–19
e-mail(s), 228
  asterisks, meaning of, in, 237
  attaching files to, 199, 233–35, 239–42, 306
  BCC function in, 235–36

downloading files from,
241–42
fact-checking, 231
open documents sent to, 199
phishing scams via, 211–12
quoting content of, 232
"reply all" option in, 214–15
size limits, 233–35
spam, 213–14, 229
temporary addresses for,
229–30
verifying accounts via, 229–30
Eraser program, 88
e-readers, 66–67
Evernote app, 325
Excel, Microsoft, 183
auto-formatting in, 187–88
automatic list maker in, 204
change-case options in, 198
copying cell content in, 193
editing cells in, 190
e-mailing open files in, 199
Format Painter tips for, 194–96
keyboard navigation tips for,
196–97
multiple spreadsheet options
in, 188
shortcuts important to, 184,
203
spell-check in, 201–2
split-screen use in, 200–201
symbol use in, 181
exposure, camera, locks for,
62–64

Facebook, 29, 163, 237
app for, 325
keyboard shortcuts for, 332–34
mail messages via, 331–32
password for, 217

privacy options on, 235,
328–29
recommendations via, 330–31
site navigation on, 253–54
file(s)
alphabetizing, 99–100
as attachments to e-mail, 199,
233–35, 239–42, 306
auto-reopening of, 115–16
backing up of, 105–6, 219–23
copying, 132–33, 152–53
deleting, 86–88, 134
locating computer, 89–90,
93–96, 241–42
moving, 132–33, 135–36,
152–53
recovery of deleted, 60, 86–87
renaming, 100, 173
saving, 93–97
storage/transfer, online, of,
105–6, 135–36, 219–23,
235, 298, 306
type-searching for, 89–92, 95
file directories
type-searching in, 89–92, 95
width adjustments in, 171–72
File History, Windows, 222–23
Final Cut program, 104
Find My Phone app, 35, 43
Firefox, 7, 156, 245–46, 251, 260,
333
privacy settings in, 264
F-key functions, Mac, 143–44
flash settings, camera, 51–54
Flickr.com, 235, 298
FlightTrack app, 316
Flipboard app, 318
Flixter app, 319
focus, camera
locks for, 62–64
shutter lag and, 50

force-quitting programs, computer, 109–10
Format Painter, 194–96
FreeConferenceCall.com, 307
free services, Internet. *See also* discount coupons/codes
  for archived Web pages, 310
  for booking travel, 270–71, 301–2
  for business reviews, 299
  for classified ads, 304–5
  for conference calling, 307
  for flight monitoring, 271, 275, 302–3
  for gas price comparisons, 305
  for large file transfers, 105–6, 235, 306
  for music streaming, 308–9
  for picture storage, 298
  for restaurant reservations, 300–301

gadgets. *See also* computers; *specific gadgets*
  camera, 4, 31–32, 50–64
  chargers for, 73
  discount codes for, 6, 72
  earbuds for, 32, 70–71
  electrical outlets for, 68–69
  e-reader, 66–67
  iPad, 8, 9, 13–14, 21–23, 43–48, 146
  iPod, 4, 8
  phone, cordless, 67–68
  phone, smart, 4, 6, 9, 12–42, 43, 62–64, 66–67, 71, 311–25, 342
  recycling of, 6
  refurbished, 5–6
  release dates for new, 4–5

tablet, 8, 9, 13–14, 21–23, 43–48, 146, 218
  terminology, basic, for, 6–9
  USB cables for, 71
GasBuddy.com, 305
Gazelle.com, 6–7
Google, 265
  as almanac, 274
  Android software by, 6–7
  business directory by, 278–79
  as calculator, 276–77
  Chromecast, 296
  conversion features by, 272, 281–82
  as dictionary, 266, 278
  Drive, 208
  Fight, 307–8
  as flight monitor, 271, 275
  image libraries via, 267–68
  image-recognition by, 268–70
  Maps app, 30–31, 312–14
  movie information via, 273–74
  package tracking via, 275
  Play, 48
  search features of, 282–84
  Street View by, 279–81
  tech support via, 104
  time-zone information via, 277
  as translator, 267
  as travel agent, 270–71
  weather information via, 274
GoogleFight.com, 307–8

hashtags, purpose of, 338–39
HealthTap app, 315
Hipmunk Web site, 301
Hulu, 295–96

iCloud, 35

Keychain, 218
icons, definition of, 8
identity theft, 211–12
image recognition software,
    268–70
images.google.com, 269
IMDB.com, 294
iMovie program, 104
InDesign program, Adobe, 189
international travel
    Airplane Modes and, 29
    Wi-Fi and, 29
Internet. *See also* browsers, Web;
    e-mail(s); free services,
    Internet; Google; social
    networks; YouTube
    ad blocking on, 260–61
    applications, phone, available
        from, 30–31, 34, 35, 43,
        66–67, 311–25, 342
    backup of computers using,
        105–6, 219–20
    copying text from, 248–49
    e-mail on, 211 15, 228 42,
        306
    fact-checking on, 231
    free services on, 297–310
    Google features on, 104,
        265–84
    identity theft via, 211–12
    logos, site, and, 253–54
    movie/TV shows on, 295–96
    online form shortcuts for,
        255–56
    passwords for, 217–18
    phishing scams on, 211 12
    pop-up menu tips for, 256–57
    printing from, 259–60
    privacy mode for browsing of,
        263–64
    scrolling tips for, 254–55

spam and, 213–14
    tech support via, 104–5
    text search within sites on,
        258–59
    videos on, 285–96
    viruses/malware on, 147–49,
        215–16
    Web browsers on, 243–64
Internet Explorer, 7
    privacy settings in, 264
iOS, definition of, 8
iPads, 146
    charging tips for, 13–14, 43,
        44–45
    iOS of, 8
    keyboard tips for, 21–23, 43,
        45
    as picture frames, 47–48
    release dates for new, 4
    rotation lock on, 46–47
    software versions of, 9
    split keyboard screen on, 45
iPhones. *See also* cameras,
    smartphone; earbuds,
    Apple; phone app(s);
    smartphones
    app deletion on, 26–27
    apps for, 30–31, 34, 35, 43,
        66–67, 311–25, 342
    autocorrect tips for, 23–25
    automatic period trick on, 13
    banner dismissal on, 38–39
    battery extension tips for,
        15–16
    battery recharging tips for,
        13–14, 43
    camera tips for, 31–32, 62–64,
        71
    directory assistance tips for, 26
    earbuds for, 32, 70–71
    e-book apps for, 66–67

international travel tips for, 29–30
iOS of, 8
keyboard tips for, 21–23, 43
list navigation on, 39
lost/stolen, 35–37
password protection on, 37, 218
QR information codes and, 34
redialing tip for, 17–18
release dates for new, 4
ring muting on, 17
screen magnification on, 27–28
screenshot retrieval on, 28
settings adjustment on, 12–18
software version of, 9
speech recognition for, 40–41
voice mail tips for, 20
wet, 18–19
Wi-Fi settings on, 32–33
iPhoto program, 234, 298
iPods
iOS of, 8
release dates for new, 4
iTunes program, 104

JotNot Pro app, 320–21

keyboards, computer
navigating open programs via, 101–2, 158, 168–70, 196–97
shortcuts, Mac, 8, 108–9, 140–42, 184, 203, 245–48, 287–88, 332–34
shortcuts, Windows, 8, 108, 160, 163, 179–80, 184, 203, 245–48, 287–88, 332–34

keyboards, smartphone/tablet
capital lock on, 21, 43
punctuation tips for, 21–23, 43
split, 45
Kindle, app for, 66–67

laptop computers. *See also* computers
password protection of, 112, 218
sleep mode use on, 77, 142–43
Libre Office program, 207
Lightroom, Adobe, 298
LinkedIn app, 325
list navigation, iPhones, 39
Lock screen, Windows, 182
lost phone apps, 35–37

Mac computers. *See also* browsers, Web; computers; iPads; iPhones; iPods
abbreviation-expander feature on, 121–22
accent/symbol insertion on, 122–24
AirDrop feature on, 135–36
auto-on/off option on, 142–43
auto-reopening of files on, 115–16
backing up, 105–6, 219–22
calculators on, 126–27
closing windows on, 133–34
data recovery on, 60, 86–87
dictation feature on, 125–26
Documents file on, 93–96
e-book apps for, 66–67
e-mail attachments on, 199, 233–35, 239–42, 306

file copying *v.* moving on, 132–33
file deleting on, 134
file saving on, 93–96
F-key functions on, 143–44
lessons available for, 145
operating systems of, 9
passwords for use of, 112, 218
PDF creation on, 136–37
Quick Look feature on, 120–21
Repair Disk Permission on, 139–40
screen activity, movies of, on, 131–32
screen magnification on, 127–28, 261–62
screenshots on, 128–30
scrolling options on, 97–98, 113–14, 254–55
shortcuts, keyboard, for, 8, 108–9, 140–42, 184
signatures, electronic, on, 118–19
speaking feature on, 116–17
system preferences on, 111
Trash Bin on, 86
Web browser tips for, 244–64
Web receipts on, 138–39
magnification, screen
on iPhones, 27–28
on Macs, 127–28, 261–62
on Windows PCs, 153–54, 261
maps.google.com, 280
megapixels, camera, 55–56
menus, definition of, 9
Microsoft, 78, 111. *See also*
Windows 8, Microsoft;
Windows PC computers
antivirus software by, 147–49
OneDrive by, 96–97

Windows 8 by, 95, 146, 148–49, 160–63, 165–70, 175–76, 215–16
Microsoft Office, 144. *See also specific programs*
Excel program in, 181, 183–84, 187–88, 190, 193–204
free alternatives to, 207–8
Outlook program in, 189, 199, 240, 242
PowerPoint program in, 181, 183–84, 189, 192, 194–96, 198–99, 201–3, 205–7
Word program in, 2, 181, 183–86, 189, 191, 194–203
Microsoft Security Essentials program, 148–49
mouse devices, 134
highlighting/selecting text with, 83–84
right-click feature of, 78–80
wheels on, 102–3
movies
information, online, for, 273–74, 319
Internet viewing of, 295–96
review sources for, 294
music player tips, computers, 103–4, 286–87

Netflix, 295–96
"new every two" slogan, 2
New York Times, 3
online, 253
Nook (e-reader)
apps for, 66–67
Notepad, date stamp in, 159–60

OneDrive feature, Windows, 96–97
OnePass program, 218
One to One lessons, Apple, 145
OpenTable, 300–301
Outlook, Microsoft, 189, 199, 240, 242
Overstrike mode, Word, 191

package tracking, online, 275
Pandora app (music), 322–23
Pandora Recovery software, 60
password(s). *See also* privacy options
  managers, 218
  for phones, 37
  for screens, home computer, 112, 149–52
  for screens, laptop, 112
  for Web sites, 217–18
PCs. *See* Windows PC computers
PDF files, creation of, 136–37
phishing, 211–12
phone app(s)
  Dictionary.com, 324
  for e-books, 66–67
  Find My Phone, 35, 43
  FlightTrack, 316
  Flipboard, 318
  Flixter, 319
  Google Maps, 30–31, 312–14
  HealthTap, 315
  JotNot Pro, 320–21
  Pandora, 322–23
  RedLaser, 34
  social media, 325, 342
  SoundHound, 317
  Spotify, 322–23
  Uber, 321–22
phones, cordless, static on, 67–68

phones, smart. *See* smartphones
photo(s)
  cropping, 61–62
  e-mailing, 233–35
  libraries, online, 267–70
  recognition software for, 268–70
  recovery of deleted, 59–60
  "selfie," 31–32, 71
Photoshop, Adobe, 8, 62, 102
Photo Slides app, 48
Picasa, 234, 298
pop-up menus, shortcut for, 256
PowerPoint, Microsoft, 183
  black-out/white-out of screens in, 205
  change-case options in, 198
  e-mailing open files in, 199
  Format Painter tips for, 194–96
  pasting text into, 189
  positioning content in, 192
  shortcuts important to, 184, 203
  slideshow presentation options in, 205–7
  spell-check in, 201–2
  symbol use in, 181
printing
  PDF creation via, 136–37
  from Web sites, 259–60
privacy options. *See also* password(s)
  for browsing, 263–64
  for computer use, 112, 149–52
  on social media sites, 235
punctuation tips, smartphone, 21–23

QR (Quick Response) information codes, 34

Quicken program, 8
Quick Look feature, Mac,
    120–21
QuickTime Player, 104, 131–32

Radio Shack, recycling by, 6
read.amazon.com, 67
Recuva program, 60
Recycling Bins. *See also* data
    recovery; Trash
    emptying of, 86
    tips for, 177–78
recycling, gadget
    drop-off locations for, 6
    online, 6
redialing numbers, 17–18
RedLaser app, 34
Repair Disk Permissions, Mac,
    139–40
"reply all" feature, e-mail, 214–15
resolution, video, 289–91
restarting computer, 77
RetailMeNot.com, 6, 72–73
reversed scrolling, computer, 114
right-clicking
    using mouse, 78–80
    using trackpad, 155–56
ring muting, smartphone, 17
RottenTomatoes.com, 294

Safari, 7, 115, 116, 260
    browser shortcuts and, 245–46
    privacy settings in, 263
    tabbed browsing in, 251–52
    text magnification in, 262
Samsung, product introductions
    by, 4
screen
    locks, 174, 182

magnification, computers,
    127–28, 153–54, 261–62
magnification, iPhones, 27–28
movies, computer, 131–32
    splitting, 45, 200–201
screenshots
    on computers, 128–30, 157
    on smartphones, 28
scrolling
    reversed, 114
    via scroll bar, 97–98
    via space bar, 254–55, 286–87
    on trackpads, 113–14
Secure Empty Trash command,
    88
"selfie" photos, 31–32, 71
self-timers, camera, 31–32, 58
SendBigFiles.com, 306
sensors, camera, 55–56
settings, smartphone, adjustment
    of, 12
shortcuts, desktop, 179–80
shortcuts, keyboard
    for Facebook, 332–34
    Mac, 8, 108–9, 140–42, 184,
        203, 245–48, 287–88,
        332–34
    for Web browsing, 245–48
    Windows, 8, 108, 160, 163,
        179–80, 184, 203, 245–
        48, 287–88, 332–34
    for YouTube, 287–88
shorthand, texting, 237–39
shutter lag, camera, 50
shutter speed, camera, 60–61
shutting down computer, 77
signatures, electronic, 118–19
Siri (speech-recognition feature),
    iPhone, 40–41
    carbud controls for, 71

sleep mode, computer, 77,
142–43
Slideshow Maker app, 48
smartphones. *See also* Android
phones; iPhones; phone
app(s)
Airplane Mode on, 14, 16
app deletion on, 26–27
apps for, 30–31, 34, 35, 43,
66–67, 311–25, 342
autocorrect tips for, 23–25
automatic period trick on, 13
background updates on, 16
banner dismissal on, 38
battery extension tips for,
15–16
battery recharging tips for,
13–14, 43
camera tips for, 31–32, 62–64,
71
data "pushing" on, 16
directory assistance for, 26
e-book apps for, 66–67
games on, 16
international travel tips for,
29–30
keyboard tips for, 21–23, 43,
45
list navigation on, 39
lost/stolen, 35–37
QR information codes and, 34
recycling of, 6
redialing tip for, 17–18
release dates for new, 4
ring muting on, 17
screen magnification on, 27–28
screenshots on, 28
settings adjustments on, 12–18
software versions of, 9
speech recognition by, 40–41
voice mail tips for, 20

wet, 18–19
Wi-Fi settings on, 16, 29,
32–33
Snake (YouTube game), 292–93
Snopes.com, 231
social networks. *See also specific
networks*
Facebook, 29, 163, 217, 235,
237, 253–54, 325,
328–34
LinkedIn, 325
Twitter, 163, 217, 237, 325,
335–42
Softtote Data Recovery Free
program, 60
SoundHound app, 317
space bar
music player controls via, 103–
4, 286–87
scrolling via, 254–55, 286–87
spam, 213–14, 229–30
Speaking command, Mac,
116–17
speech-recognition
on iPhone, 40–41
on PC computers, 165–68
spell check, 201–2
split-screen options, 45, 200–201
Spotify, 308–9
app for, 322–23
Start menu, Windows
hidden, 162–63
navigation in, 163, 168–70
restoration of classic, 161–62
shortcut for, 160
Stellar Phoenix program, 87
Stickies, Windows, 116
symbols, typing of, 122–24,
180–81
System Restore, Windows,
224–25

Tab key, online forms and,
  255–56
tablets, 218
  charging tips for, 13–14,
    44–45
  keyboard tips for, 21–23, 43,
    45
  as picture frame, 47–48
  rotation lock on, 46–47
  smartphone similarities to, 43
tabs, browser, 251–52
tech companies
  business models of, 1–3
  "new every two" slogan by, 2
  new product releases by, 2, 4–5
tech support, Internet, 104–5
TED conferences, 3
10minute-mail.com, 230
text
  accents/symbols, 122–24,
    180–81
  auto-formatting of, 184–86
  change-case tip for, 198
  date stamps for, 159–60
  deletion, 84–85, 210–11
  dictation, 125–26, 165–68
  in Excel cells, 190, 193, 204
  magnification, 153–54, 261–62
  Overstrike mode for, 191
  pasting without formatting,
    189
  replacing, 84–85
  selection of, 83–84, 92–93,
    248–49
  shorthand, 237–39
  type-searching for, 258–59
TextEdit, Windows, 116
texting, shorthand for, 237–39
Time Machine program, Apple,
  220–22
time zone information, Web

  resources for, 277
T-Mobile, 29
trackpads
  right-clicking on, 155–56
  scrolling on, 113–14
translation, online, 267
Trash. See also data recovery;
    Recycling Bins
  emptying of, 86
travel
  booking, online, of, 270–71
  monitoring, online, of, 271,
    275, 302–3, 316
  smartphone tips for
    international, 29–30
TripAdvisor.com, 299
tripods
  lamps as, 56
  strings as, 57
Twitter, 163, 237, 335
  apps for, 325, 342
  following users on, 336
  hashtags, purpose of, on,
    338–39
  password for, 217
  private messaging on, 341
  symbol guide for, 337–38
  visibility of posts to, 340–41
type-searching
  of files in directories, 89–92,
    95
  of pop-up menus, 256–57
  of text on Web sites, 258–59
  of tiles on Start screens, PCs,
    95, 163

Uber app, 321–22
"Undo" command, computers,
  80, 211
URL addresses, 244–45, 247–48

shortening of, 249–50
USB cables, 71

videos. *See* YouTube
Vimeo, 104
viruses/malware, 147–49, 215–16
voice mail, bypassing of, 20

weather monitoring, online, 274
Web browser. *See* browsers, Web
Web receipts, 138–39
wet phones, salvaging of, 18–19
Wi-Fi features, smartphone
battery life and, 16
hot-spot announcements and,
32–33
international travel and, 29
Wi-Fi, phone interference and,
67–68
Windows 8, Microsoft, 146
antivirus software for, 148–49,
215–16
navigation between screens in,
168–70
Speech Recognition software
in, 165–68
Start menus and, 160–63,
168–70
type-searching in, 95, 163
Windows-logo key functions
in, 175–76
Windows Defender program,
148–49
Windows Live Mail program, 242
Windows-logo key, functions of,
175–76
Windows Media Player, 104
Windows PC computers. *See also*
browsers, Web; computers;

Start menu, Windows;
Windows 8, Microsoft
antivirus software on, 147–49,
215–16
backing up, 105–6, 219,
222–23
Backspace key functions on,
147, 156
closing windows on, 158
cursor size options on, 164–65
data recovery on, 60, 86–87
date stamps on, 159–60
declining use of, 12
Delete key on, 147
Documents file on, 93–96
e-book apps for, 66–67
e-mail attachments on, 233–
35, 239–42
file copying on, 152–53
file directory options on, 89–
92, 95, 171–72
file renaming on, 173
file saving on, 93–97
Notepad features on, 159–60
OneDrive for, 96–97
password protection of, 149–52
Recycling Bin on, 86, 177–78
right-clicking on, 155–56
screen locking on, 174, 182
screenshots on, 157
screen/text magnification on,
153–54, 261
shortcuts, desktop, on, 179–80
shortcuts, keyboard, for, 8,
108, 160, 163, 179–80,
184
software versions of, 9
start menus on, 160–63
symbol insertion options on,
180–81

System Restore feature on,
224–25
tile-desktop navigation on,
168–70
type-searching on, 95, 163
voice-command options on,
165–68
Web browser tips for, 244–64
Windows-logo key functions
on, 175–76
Windows Speech Recognition,
165–68
Word, Microsoft, 183
auto-formatting in, 184–86
change-case options in, 198
e-mailing open files in, 199
features of, 2
Format Painter tips for, 194–96
keyboard navigation tips for,
196–97
Overstrike mode in, 191
pasting text into, 189
shortcuts important to, 184,
203
spell check in, 201–2
split-screen in, 200–201
symbol use in, 181

statistics on use of, 285
tech support via, 105
video resizing on, 288–89
video sharpening on, 289–91

ZoneAlarm Free program, 149

Yahoo, 253
Yelp.com, 299
app for, 325
"Yes" and "No" keys, computer,
81–82
YouTube, 309
keyboard shortcuts for, 287–88
search tips for, 291–92
secret game on, 292–93
site navigation on, 253–54
space bar functions on, 104,
286–87